The Elementary Language of Solid State Physics

The Elementary Language of Solid State Physics

M. H. B. STIDDARD

Department of Chemistry
University College, London

ACADEMIC PRESS · 1975
LONDON · NEW YORK · SAN FRANCISCO
A Subsidiary of Harcourt Brace Jovanovich, Publishers

ACADEMIC PRESS INC. (LONDON) LTD
24/28 Oval Road
London NW1

United States Edition published by
ACADEMIC PRESS INC.
111 Fifth Avenue
New York, New York 10003

Library of Congress Catalog Number: 75-19677
ISBN: 0-12-671050-3

FILM SET BY COMPOSITION HOUSE, SALISBURY, ENGLAND

PRINTED IN GREAT BRITAIN BY WHITSTABLE LITHO LTD., KENT

Preface

This book was written primarily as an aid to students of chemistry who wish to learn something about solid state physics. Its writing was further motivated by the author's feeling that solid state physics is far too interesting to be left to physicists! It is also directed towards the person who may find that a standard introductory text such as that of Kittel is a bit hard-going but nevertheless finds that the non-mathematical introductory books are too trivial for serious study.

In this book an attempt has been made to bridge the gap between the linguistic background of the typical chemist, who has some experience of elementary physics, crystallography, group theory, and quantum mechanics, and that required to benefit from 'physicists' solid state physics' books. Indeed this is *not* a complete text book on solid state physics. Purposely it has been kept very short and consequently some topics have been omitted. Aspects of potential interest to chemists have been selected particularly, but a genuine attempt has been made to acquaint the reader with most of the elementary language of the subject. No apologies are made for emphasis of the term *language*. It is the author's personal feeling that many people experience a repulsion from an interesting topic because of language problems. However, since a working knowledge of a phrase-book containing current slogans (easily assimilated) cannot be equated with even an elementary knowledge of the language, quite a lot of effort is required.

Solid state physics as an identifiable discipline, although lacking an obvious inception, was probably born during the first years of this century. However, the crystalline state, aspects of which constitute basic solid state physics, had attracted attention for centuries both from a scientific and aesthetic standpoint. The Greeks first gave the name *Κρύσταλλος*, literally meaning frozen ice, to the mineral quartz because they believed it was water frozen by intense cold, a belief which survived into medieval times. As Pliny stated in his *Natural History* 'crystal is only found in those high places where the winter snows have gathered in great quantity, and it is surely ice; and for this reason the Greeks have given it this name'. Since the two distinguishing features of crystal were its transparency and geometric regularity, it was argued that all materials with these properties were forms of crystal, so that the word took on its present meaning.

Serious study of crystals was begun in the seventeenth century by Stensen and by Guglielmini and their observations were rationalised into a physical law by Romé de l'Isle in 1772: 'In all crystals of the same substance, the angles between corresponding faces have the same value'. But the first great observer of crystals whose name has become established in the folklore of crystallography was undoubtedly René-Just Hoüy, the son of a poor French weaver. Popular belief is that he accidentally dropped some calcite crystals and that he noticed the similarly shaped fragments. From this observation he concluded that crystals were built up from a large number of identical units. The natural momentum of interest in crystals continued into the nineteenth century and on which were laid the foundations of modern mathematical crystallography. Observations of external morphology were transformed into a mathematical language by a number of workers; the names of Bravais, Sohncke, Fedorov, Schoenflies and others come to mind.

Thus to 1900 which not only marked the beginning of this century but also the beginning of quantum physics. It would be an impossible task to mention all the major contributions to solid state physics since 1900, but clearly there are a few of pivotal importance.* The first application of the new quantum theory to an atomic system was effected by Einstein in 1907 in his famous work on the specific heat of monatomic solids. In 1912, three major papers were published. Debye made further contributions to the theory of specific heats, an area of study still of considerable interest. The classical work of Born and von Kármán of the same year on lattice vibrations had far-reaching consequences, but perhaps the work of von Laue, Friedrich, and Knipping on the diffraction of X-rays must take pride of place. The explosive progress in formal quantum mechanics in the 1920s naturally resulted in successful application to the electrons in solids: in one year, 1928, we find two papers (by Sommerfeld and by Bloch) on the free electron theory of metals (based on the newly discovered Fermi–Dirac statistics) and the quantum mechanics of electrons in crystal potentials respectively. Most of our attention will be directed towards contributions to our understanding of solid state properties made during this fruitful period since one joy of elementary solid state physics is that tangible progress can be made by application of very simple ideas: by describing mathematically a very simple model and examining the consequences thereof.

It is hoped that the average serious student of physical chemistry who may read this book will agree that the level at which, for example, quantum mechanics is employed is *easily* within his or her grasp. Although elementary, this book does not conform with the highest ideal of some educationalists that nothing too difficult should be tackled. The mathematical content is

* A review by J. C. Slater 'Interaction of Waves in Crystals' (*Rev. Mod. Phys.* **30**, 197 (1958)) contains a very useful list of important papers.

very simple, although some familiarity with differential equations, vectors, and matrices is assumed. The contents follow a conventional order. Semi-classical considerations of ionic crystals are followed by aspects of crystal symmetry and diffraction theory. Subsequently the electron theory of solids is developed and finally we conclude with a short discussion of defects. Detailed lists of references are not included since they seem inappropriate in an elementary book of this type. However, mention is made of a few papers which appear of particular historical and/or scientific importance. The reader should not shun the reading of early literature. The physics of 1900–1930 is far more interesting than the chemistry of the same period! At the end of each chapter the reader may be referred to sources which contain more intensive or extensive details of the material dealt with in that chapter. Furthermore, mention may be made at this stage of general books which the author has found very useful and which may be mentioned by name subsequently in the text.

General Solid State Physics

Introduction to Solid State Physics, 4th edition by C. Kittel (John Wiley, 1971)
Modern Theory of Solids by F. Seitz (McGraw-Hill, 1940)
Solid State Physics by A. J. Dekker (1957; now available as Macmillan Student Edition)
An Introduction to the Quantum Chemistry of Solids by C. M. Quinn (Clarendon Press, Oxford, 1973)
The Theory and Properties of Metals and Alloys by N. F. Mott & H. Jones (1936; now available from Dover Publications, New York)

Crystallography

An Introduction to Crystallography, 4th edition by F. C. Phillips (Oliver and Boyd, Edinburgh, 1971)
X-ray Crystallography by H. J. Buerger (John Wiley, 1962)

Statistical Mechanics

An Introduction to Statistical Thermodynamics by T. L. Hill (Addison-Wesley, 1960)
Statistical Mechanics by J. E. Mayer and M. G. Mayer (John Wiley, 1940)

Quantum Mechanics

Quantum Mechanics by L. I. Schiff (McGraw-Hill, 1955)

A number of other books could be recommended but let it suffice to mention one or two on general topics. For general physics written with sympathy and authority, *The Feyman Lectures* enjoy a position of high status (although the introduction to quantum mechanics may be considered idiosyncratic). With chemists in mind, a very attractive introduction to chemical physics may be found in *Molecular Structure, the Physical Approach* by J. C. D. Brand and J. C. Speakman (Edward Arnold, 1960). For those interested in the development of quantum mechanics (at the popular level) George Gamow's *Thirty Years that Shook Physics* (Heinemann, 1972) makes an entertaining bedside companion. For more serious scientific biography *Einstein, the Life and Times* by R. W. Clark (Hodder and Stoughton, 1973) is worthy of attention. It is also a source of numerous references. As an inspiration to anyone seriously interested in chemical physics, who could be a better mentor than Irving Langmuir, whose biography *The Quintessence of Irving Langmuir* by A. Rosenfeld (Pergamon Press, 1966) makes fascinating reading. Finally Karl Popper's *Conjectures and Refutations* (Routledge and Kegal Paul 1963) should be *required* reading for all chemists!

It is hoped above all that the student will benefit from reading the following pages *ab ovo usque ad mala*. The book has been written to inform rather than to impress. Indeed, if the student's reaction on completion is that it is insufficiently demanding and that Ziman's books are more appropriate, then it will have achieved its aim.

Finally, a word of advice by way of a quotation from Silvanus Thompson's famous book *Calculus Made Easy*: 'What one fool can do another can'. This advice has acted as a frequent stimulus to the author in moments of depression.

London M. H. B. Stiddard
March 1975

Acknowledgements

It is a pleasure to acknowledge the encouragement and support of my family in the writing of this book, during leave of absence from University College, London. The hospitality of staff and students at Imperial College, London was appreciated during this period.

Parts of this book have been included in undergraduate lecture courses and critical comments from physics and chemistry students have proved very helpful. Various colleagues have made general comments about the contents of this book, but I am indebted, in particular, to Alan Knapp (Mullard Research Laboratories, Redhill) who read the manuscript in its entirety and to 'Thiru' Thirunamachandran (Chemistry Department, University College) who read selected chapters. Their constructive comments and help in removing ambiguities and errors have proved invaluable. Needless to say, remaining errors are my sole responsibility.*

Finally I record my gratitude to Maruschka Malacos and John Cresswell (Chemistry Department, University College) respectively for typing the manuscript and preparing the figures.

M. H. B. S.

* Me adsum qui feci in me convertite ferrum! (Virgil, *Aeneid*)

Contents

Symbols and Physical Constants

Where possible conventional symbols are employed although this has meant an over-utilisation of certain Greek letters; however no confusion should result from this repetition. An attempt has been made to use SI units although electronvolts and ångström units still offer some convenience and are used occasionally. Common physical constants* which the serious student may require are given for convenience in the following list.

Physical constant	*Symbol*	*Magnitude*
mass of an electron	m	$9{\cdot}108 \times 10^{-31}$ kg
1/12 of mass of C^{12}	AMU	$1{\cdot}6604 \times 10^{-27}$ kg
1 electronvolt	eV	$1{\cdot}6022 \times 10^{-19}$ J
Planck constant	h	$6{\cdot}6262 \times 10^{-34}$ J s
$h/2\pi$	$\hbar$	$1{\cdot}0546 \times 10^{-34}$ J s
Avogadro number	N_0	$6{\cdot}0222 \times 10^{23}$ mol^{-1}
Boltzmann constant	k_B	$1{\cdot}3806 \times 10^{-23}$ J K^{-1}
gas constant	R	$8{\cdot}3143$ J K^{-1} mol^{-1}
charge on one electron	e	$1{\cdot}6022 \times 10^{-19}$ C
velocity of light *in vacuo*	c	$2{\cdot}9979 \times 10^{8}$ m s^{-1}
ångström unit	Å	10^{-10} m
Bohr magneton	μ_B	$9{\cdot}2741 \times 10^{-24}$ m^2 A
permeability of a vacuum	μ_0	$4\pi \times 10^{-7}$ m kg s^{-2} A^{-2}
permittivity of a vacuum	ε_0	$8{\cdot}8542 \times 10^{-12}$ m^{-3} kg^{-1} s^4 A^2

* Numero deus impare gaudet! (Virgil, *Eclogues*)

1

Structure of Ionic Solids

By way of an introduction to solid state physics, let us begin by discussing a topic with which everyone is familiar and at the same time perhaps uncover some hidden problems. Chemists particularly will be very familiar with, or perhaps have experienced a surfeited use of, the word structure. For example, where *inorganic* chemistry is concerned, an emphasis on the structure of compounds rather than their reactions now appears to be the norm. But what do we mean by structure in the context of solid state studies? First and self-evidently, we imply some knowledge of the disposition of the constituent atoms of the solid, but second it is common to infer that structure includes some indication of the model which most closely describes the interactions between the atoms and provides some rationale for their stereochemical relationship. In order to explore this concept of structure, we consider one of the simplest systems, that of ionic solids.

1.1 Ions and ionic size

Although we may regard ionic solids as 'simple' this does not imply there are no difficulties! Indeed, we may ask immediately the non-trivial question: what is the evidence that crystalline sodium chloride contains the ions Na^+ and Cl^-? Any conditioned response which relies on the significance of the physical properties of solutions or melts as an indication of the constitution of the solid is clearly inadmissible as is any response which includes the oneiric term electronegativity or the Born–Haber cycle. We are *not* going to attempt to answer this thorny question; we shall refer to it again, but really prefer to allow the curious reader the opportunity to consider it for himself or herself. At this stage we *assume* the model that a crystal of sodium chloride contains a regular array of Na^+ and Cl^- ions whose separation may be determined by X-ray crystallography (see Fig. 1.1).

We now face a further thorny question. It is assumed in the elementary theory of the solid state that it is a realistic procedure to apportion the Na^+—Cl^- separation in sodium chloride, for example, into contributions from individual spherical ions, so that *ionic sizes* may be assigned. Lists of ionic sizes frequently appear in chemistry and physics text-books as 'facts' without even mention of the inherent assumptions. It is a worthwhile exercise to examine the concept of ionic size in more detail. This will be our first task.

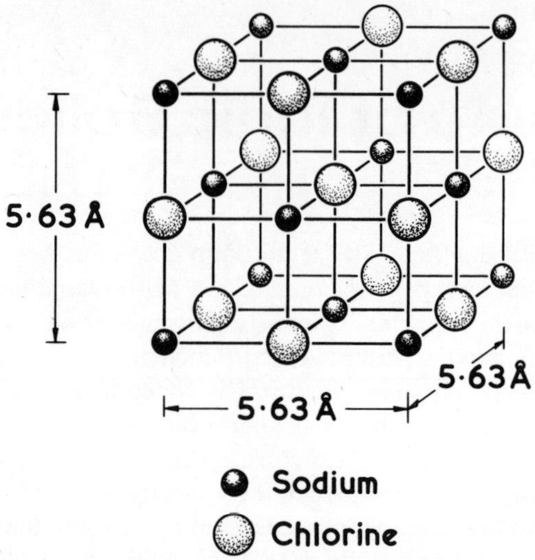

FIG. 1.1 Structure of the sodium chloride crystal.

The most compelling factual evidence that one might assign, in principle, sizes to individual ions in structurally related compounds is the approximate self-consistency of crystallographic data. That is, if in a family of binary compounds $A_i^+B_j^-$, the separation $A_i^+—B_j^-(r_{ij})$ is assumed to be the sum of the contribution of the ions $[r^+(A_i) + r^-(B_j)]$ then differences in apparent ionic radii Δr_i^+ or Δr_j^- are roughly constant. For example, in Table 1.1 values of Δr_j^- calculated from the nearest neighbour separations in sodium, potassium, and rubidium halides are presented. Immediately we may be sensitive to the magnitude of the approximation. The values of Δr_j^- may be taken as approximately constant but hardly justify the profound belief in the absolute constancy of ionic radii sometimes displayed. However, transferability of data is very important in physical science and a critical quantitative use of ionic radii is acceptable. But even assuming an additivity rule for ionic radii, absolute values of the latter cannot be calculated directly without making some assumptions whereby the absolute size of at least one ion may be fixed.

TABLE 1.1 Δr_j^- values calculated from ionic separations in alkali metal halides

Cation	$[r^-(\mathrm{Cl}) - r^-(\mathrm{F})]$ Å	$[r^-(\mathrm{Br}) - r^-(\mathrm{Cl})]$ Å	$[r^-(\mathrm{I}) - r^-(\mathrm{Br})]$ Å
Na	0·503	0·179	0·248
K	0·473	0·151	0·235
Rb	0·476	0·154	0·226

It was originally proposed by Landé that in lithium iodide, the lattice parameter is determined by contact between the iodide ions; hence the I^-—I^- nearest neighbour distance should be twice $r^-(I)$. This procedure, which leads to a value of 2·12 Å for $r^-(I)$, was also followed by Bragg except that he chose to assume that silicates contain O^{2-} ions in contact. A value of 1·35 Å for $r^{2-}(O)$ is thus calculated. The simplicity of this approach is very attractive but for no *obvious* reason the preferred technique is to relate the radius ratio of isoelectronic cations and anions to their polarisabilities. This is not very satisfactory since, of course, free ion polarisabilities cannot be measured. However, refractive index measurements coupled with the assumption that the proton has zero polarisability can lead to apparent values of ionic polarisabilities. Now the simplest way in which the size of an ion can be related to its polarisability is to assume a hydrogenic behaviour whereby each outer electron moves in an effective field due to the nucleus and the other electrons; in other words the outer electrons are influenced by an *effective nuclear charge* Z^* which may be related to the nuclear charge Z_n by the equation

$$Z^* = Z_n - \sigma \tag{1.1}$$

(where σ is a screening parameter).

According to this approximation, the ratio of the radii of the outer electrons (assumed to be the ratio of the ionic radii) of isoelectronic ions is equal to the inverse ratio of the corresponding effective nuclear charges. The ratio of the polarisabilities is equal to the inverse ratio of the fourth power of the effective nuclear charges, so that there is fourth root relationship between polarisability ratio and radius ratio. This approach was first used by Wassastjerna and values for $r^-(F)$ and $r^{2-}(O)$ so obtained were used by Goldschmidt in a compilation of a large number of ionic radii and with some modification by Zachariasen.

The approach of Pauling[1] was related to that of Wassastjerna except that the ratio of the effective nuclear charges of the isoelectronic ions was calculated to give absolute sizes without direct reference to measured ionic polarisabilities. One aspect of the Pauling approach which is distinct, however, is that different effective nuclear charges (or different screening parameters) are assumed for different atom properties. In fact, the *polarisability* screening parameters for the outer *s* and *p* electrons in the rare gas atoms Ne → Xe were first calculated from experimental atom polarisabilities. These screening parameters were then modified appropriately for the corresponding isoelectronic alkali metal and halide ions. Minor empirical corrections were made to some of these on the basis of a comparison with experimental data in solution, and finally conversion to *size* screening parameters was effected (see Table 1.2). These may finally be employed for the

TABLE 1.2 Size screening parameters for outer electrons in alkali metal and halide ions

F^-	Na^+	Cl^-	K^+	Br^-	Rb^+	I^-	Cs^+
4·52	4·52	10·94	10·80	27·08	26·58	42·29	41·31

calculation of absolute ionic sizes: e.g. for KCl we have

$$r^+(\text{K}) + r^+(\text{Cl}) = 3{\cdot}147\ \text{Å}$$

and

$$\frac{r^+(\text{K})}{r^-(\text{Cl})} = \frac{17 - 10{\cdot}94}{19 - 10{\cdot}80}$$

Hence $r^+(\text{K}) = 1{\cdot}34$ Å and $r^-(\text{Cl}) = 1{\cdot}81$ Å.

Pauling determined the radii of the alkaline-earth and chalcogen ions from those of the alkali metal and halide ions in two stages. First a 'univalent' radius of the divalent ion was defined as the radius that the ion would have in a fictitious NaCl-type crystal in which the Coulomb interactions have a magnitude corresponding to ionic charges of unity. The ratio of the 'univalent' radius of the divalent ion and the ionic radius of the isoelectronic alkali metal or halide ion was taken to be equal to the inverse ratio of the corresponding effective nuclear charges. The 'univalent' radius was then transformed into the 'real' divalent radius by means of a factor which follows readily from the Born model (see below).

No apologies are offered for taking the reader through the labyrinth leading to Pauling radii. The uncertain assumptions serve to illustrate that quoted ionic radii cannot have any absolute significance. The values normally refer to a NaCl-type lattice and modifications must be made for other stereochemical arrangements. At best we have a transferable parameter of *some* quantitative significance, but its use as a basis for discussing, for example, 'bond-shortening' due to covalent contributions cannot be accepted.

1.2 Close packing of ions and simple crystal structures

The basic feature of the model we have used so far in discussing ionic size is that the ions may be viewed as charged spheres in contact. It is of interest to investigate the utility of this simple idea in discussing the atomic arrangement in crystals. For simplicity, let us consider a fictitious ionic solid in which all the ions are identical! We may think of our solid as a metallic crystal consisting of positive ions embedded in an electron continuum (the jellium model in fact). What arrangements of the ions are possible? We may consider this question by building a 'hard sphere' structure layer by layer. The first layer of spheres may be most efficiently packed in the hexagonal arrangement shown in Fig. 1.2(a). With respect to axes perpendicular to this plane the centres of each of the spheres are at positions labelled A; interstices between the spheres are labelled B and C. We consider the addition of a second close-packed layer and it is immediately clear that the centres of the spheres of this

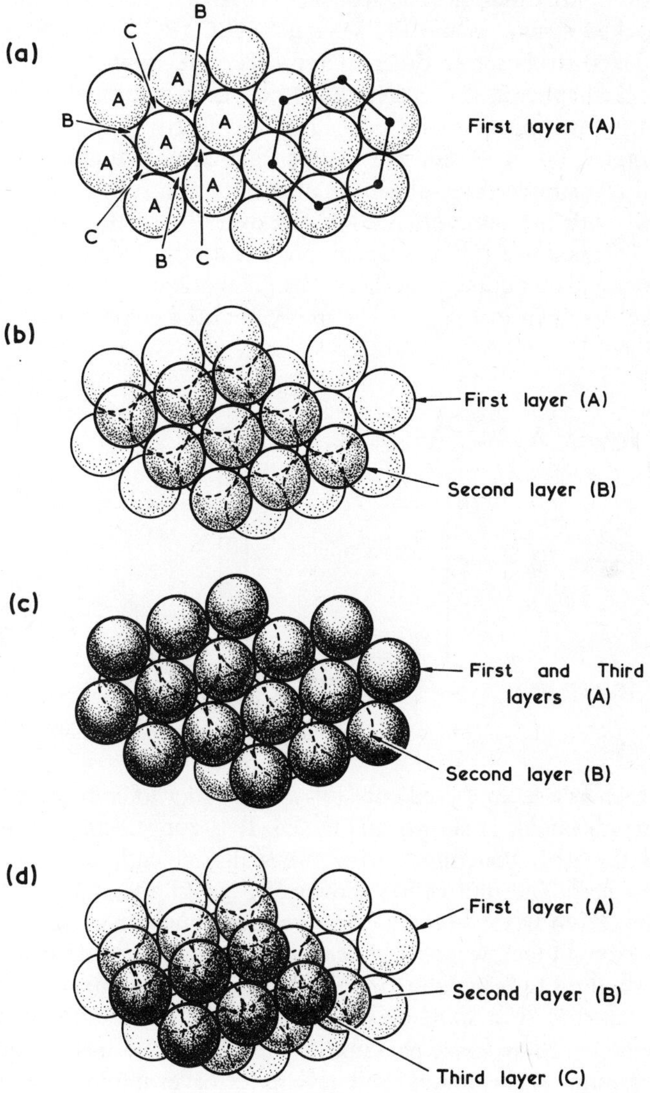

FIG. 1.2 Close packing of spheres.

layer may coincide with *B or C*. In Fig. 1.2(b) we choose the former. An interesting situation now arises on addition of the third layer, since different structures arise if we choose that the centres of the spheres of this layer coincide with either *A or C* (see Figs 1.2(c) and (d)). In both cases the packing efficiency is identical (74% of the total volume is occupied by the spheres) but the structures resulting from the regular sequences *ABABA*... and *ABCABC*... are different. The sequence *ABABA*... results in the *hexagonal* close-packed structure and the sequence *ABCABC*... results in the *cubic* close-packed structure. In order to appreciate the *symmetry* of the arrays of close-packed spheres, it is convenient to choose some small repeat element of the structure known as the *unit cell*. We shall return to this in more detail in Chapter 2, but it is sufficient at this stage to note qualitatively that the unit cell is a simple three-dimensional unit chosen so that if repeated contiguously with the same orientation, the overall structure would be reproduced. This is not a *rigorous* definition, but it allows us to *describe* simply our hexagonal and cubic structures. The prismatic unit cell of the hexagonal structure is shown in Fig. 1.3. The unit cell of the cubic structure, which is

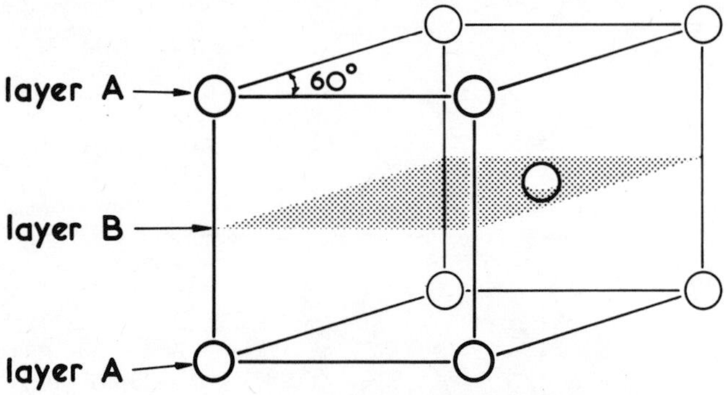

FIG. 1.3 Hexagonal unit cell (spheres reduced for clarity).

best chosen as a face centred cube (or *fcc*) in order to emphasise the overall cubic arrangement, is shown in Fig. 1.4. It is convenient at this point, to mention the other structure to which we shall frequently refer: body centred cubic (or *bcc*). The unit cell is shown in Fig. 1.5 and must be firmly distinguished from hexagonal and *fcc* structures since no close-packed layers are involved. In fact, whereas *twelve* nearest neighbours are involved in the latter structures only *eight* are involved in the *bcc* structure.

The structure of a number of elements may be described in terms of hexagonal or cubic close-packing. Of the common metals, magnesium, zinc, titanium, β-chromium, and rhenium, for example, have hexagonal structures while α-calcium, aluminium, lead, γ-iron, and nickel, for example,

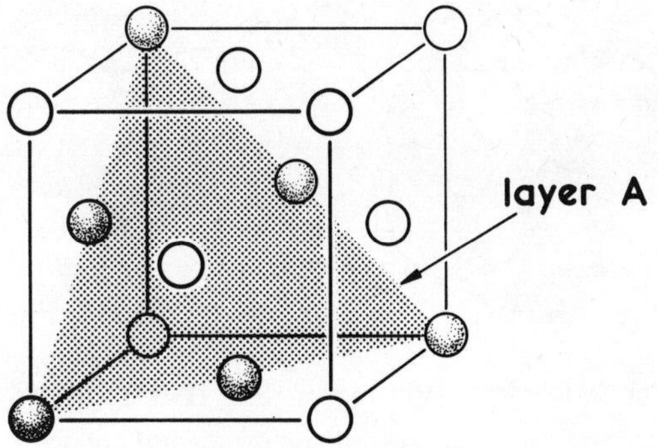

FIG. 1.4 Face centred cubic unit cell (spheres reduced for clarity).

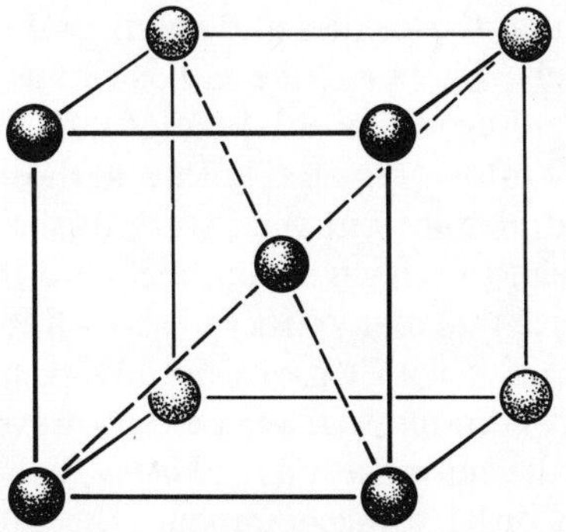

FIG. 1.5 Body centred cubic unit cell (spheres reduced for clarity).

have *fcc* structures. Examples of metals with *bcc* structures are the alkali metals, α-chromium, tungsten and tantalum.

On returning to the problem of the arrangement of ions in crystals, let us inspect Fig. 1.2(b). We note that the holes in the second layer fall either over holes or over centres of spheres of the first layer. In the former case an *octahedral hole* is formed in the latter case a *tetrahedral hole* is formed between the first and second layers. A magnified view is given in Fig. 1.6, and some further thought (or inspection of models, if necessary) will reveal that there are twice as many tetrahedral holes as there are octahedral holes in a close-packed structure. The existence of holes gives a clue to the model chosen to describe a number of inorganic structures: the anions may be considered to constitute a close-packed array in which tetrahedral and/or octahedral holes are occupied by cations. Immediately it will be recalled that this model is identical in essence with that suggested by Landé as a means of fixing ionic radii. By calculating the maximum radii of spheres which would fit into octahedral and tetrahedral holes we might qualitatively glean something of

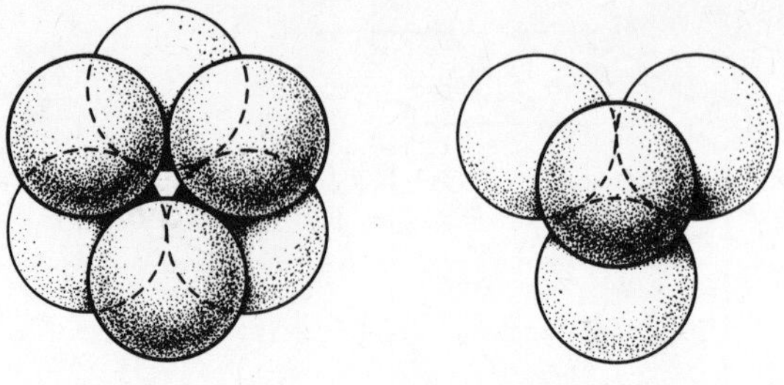

FIG. 1.6 Tetrahedral and octahedral holes.

the appropriateness of this model. Simple geometry, which can be confirmed by the reader, shows that tetrahedral and octahedral hole 'radii' are 0·225 r and 0·414 r respectively, where r is the radius of the close-packed spheres. Thus, assuming the Pauling radius of 1·40 Å for the O^{2-} ion, cations smaller than 0·315 Å should fit into tetrahedral holes and those smaller than 0·580 Å should fit into octahedral holes without disturbing the close-packed array. It would be wrong to read too much significance into these numbers and the reader will soon discover that many structures which are described in terms of close-packed anion arrays with cations occupying holes do not conform with these geometric conditions. In many cases, however, it is convenient and profitable to consider the anion array as *expanded* slightly beyond the normal contact separation in order to accommodate the cation. For example, in crystalline sodium chloride, we may consider that the Cl^- ions constitute a cubic close-packed array with Na^+ ions occupying octahedral sites. *This description is one of convenience rather than reality* and the structure of other binary compounds may be similarly rationalised. Thus, for example, in the zincblende (ZnS) structure it is apposite to assume that the S^{2-} ions form an *fcc* array while the Zn^{2+} ions occupy alternate tetrahedral holes. Table 1.3 contains details of structures *derived* from cubic close-packing and it is desirable that the serious reader should spend some time familiarising himself or herself with details of these structures, which are dealt with in the standard chemistry text-books.

1.3 Binding energy of ionic crystals

The foundation of the classical electrostatic theory of binding in ionic solids was laid in the early part of this century by Born and by Madelung. The basic concept is a simple one: the electrostatic attraction of spherical ions in a crystal is balanced by a repulsive interaction to give a stable atomic arrangement with a net binding energy relative to free ions. According to

TABLE 1.3 Structures derived from cubic close-packing

Structure type	*Holes used*	*Fraction occupied*	*Examples*
Rock salt	octahedral	1/1	Halides of Li, Na, K, Rb; NH_4Cl AgF, AgCl, AgBr; MgO, CaO, SrO, BaO, TiO, VO, MnO
Zincblende	tetrahedral	1/2	ZnS, CuCl, CuBr, CuI, AgI, BeS, β CdS, CdTe, GaP, GaSb
Fluorite	tetrahedral	1/1	CaF_2, SrF_2, BaF_2, PbF_2; UO_2, NpO_2
Antifluorite	tetrahedral	1/1	Li_2O, Na_2O, K_2O, Rb_2O

Gauss' law, we may treat a charged sphere as a point at which the charge is concentrated. This simplification makes the calculation of the Coulombic part of the potential fairly straightforward, but since the finite size of the ions is responsible for the repulsions, the choice of a short range force field is less easy. But let us deal separately with these two contributions to the potential.

The Coulombic potential

According to Coulomb's law, the force F between two point charges $\pm q$ separated by a distance r is given by

$$F = \frac{\pm q^2}{4\pi\varepsilon_0 r^2} \tag{1.2}$$

The electrostatic interaction (that is the work done in bringing together the charges from a separation of ∞ to r) is

$$-\int_{\infty}^{r} F \,\mathrm{d}r = \frac{\pm q^2}{4\pi\varepsilon_0 r} \tag{1.3}$$

Now if Φ_{ij} is the electrostatic interaction energy between the ions i and j, the total energy of any *one* ion, Φ_i is given by

$$\Phi_i = \sum_j{}' \Phi_{ij} \tag{1.4}$$

$$= \sum_j{}' \frac{\pm q^2}{4\pi\varepsilon_0 r_{ij}} \tag{1.5}$$

in which the summation includes all ions except $i = j$.

Now r_{ij} may be expressed in terms of the nearest neighbour separation r_0 such that

$$r_{ij} = p_{ij} r_0 \tag{1.6}$$

Hence

$$\Phi_i = \sum_j{}' \frac{\pm q^2}{4\pi\varepsilon_0 p_{ij} r_0} \tag{1.7}$$

$$= \frac{-Aq^2}{4\pi\varepsilon_0 r_0} \tag{1.8}$$

Or for 1 mole (2 N_0 ions),

$$\Phi = \frac{-N_0 A q^2}{4\pi\varepsilon_0 r_0} \tag{1.9}$$

where

$$A = \sum_j{}' \frac{\pm 1}{p_{ij}} = \text{Madelung constant.}$$

(Note the choice of sign convention whereby the attractive components in the Madelung constant are given a positive sign and *vice versa*).

Clearly A will be a constant dependent on the crystal structure and its evaluation is of sufficient importance for a separate discussion.

Evaluation of Madelung constants[2]

As a simple example, let us compute the Madelung constant for an infinite linear chain of ions of alternating unit charge at a separation r_0. If we use a negative ion as reference, our sign convention demands a positive sign to be associated with cations and a negative sign to be associated with anions in the chain.

We see immediately

$$A = 2\left(\frac{1}{1} - \frac{1}{2} + \frac{1}{3} - \frac{1}{4}\right)$$

and noting that

$$\ln(1 + x) = x - \frac{x^2}{2} + \frac{x^3}{3} - \frac{x^4}{4} \cdots$$

it follows that

$$A = 2 \ln 2$$

Clearly we have chosen a very simple (if not very useful) example whereby the evaluation of A is straightforward. Difficulties arise, however, when we tackle a three-dimensional case. For example, let us consider sodium chloride (see Fig. 1.1) which we may treat similarly to the one-dimensional chain by noting that with reference to any negative ion there are 6 nearest neighbours (positive), 12 next-nearest neighbours (negative), 8 next-next-nearest neighbours (positive) etc. Calculation of the distances between the ions in terms of the nearest neighbour separation is straightforward and we thus obtain for the Madelung constant:

$$A = \frac{6}{1} - \frac{12}{2^{1/2}} + \frac{8}{3^{1/2}} - \frac{6}{2} + \frac{24}{5^{1/2}} - \cdots$$

$$= 6{\cdot}000 - 8{\cdot}485 + 4{\cdot}620 - 3{\cdot}000 + 10{\cdot}733$$

It is immediately apparent that the convergence is poor and a more subtle treatment is necessary. An elegant method for such summations was developed by Ewald[3] but it would not be very profitable to give details here. However, when the interested reader is more familiar with Fourier expansions and reciprocal lattice space, examination of Ewald's paper would be worthwhile. Instead, we will limit our discussion to an alternative approach which involves groups of ions chosen so that the group has an overall electric charge of zero, where necessary dividing an ion between groups by assigning to it a fractional charge. As an elementary illustration of this technique, we may consider the square array shown in Fig. 1.7, which might be a section through a sodium chloride crystal. The first approximation is obtained by considering the square group of ions surrounding the reference ion. Here we

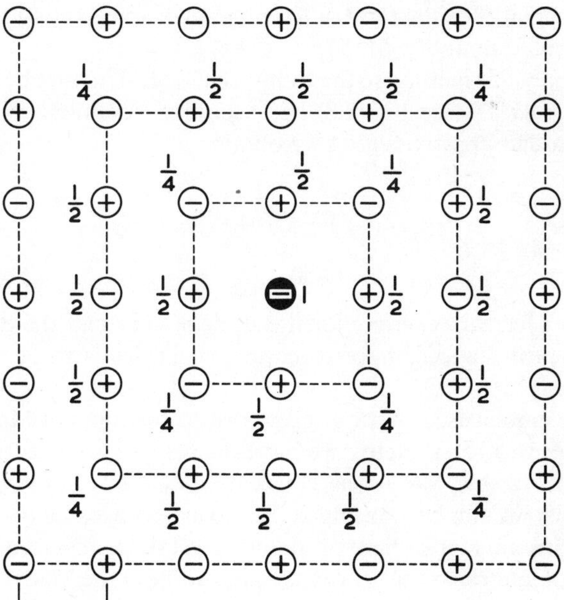

FIG. 1.7 Calculation of the Madelung constant for a square array of ions.

must ascribe fractional charges as indicated in the diagram (charge on the ions at the middle of the edges is shared by *two* cells and that of the ions at the corners is shared by *four* cells).

Hence to a first approximation:

$$A = \frac{4 \cdot \frac{1}{2}}{1} - \frac{4 \cdot \frac{1}{4}}{2^{1/2}}$$

$$= 1{\cdot}2929$$

The second approximation is obtained by considering ions within the next square array. Again fractional charges are associated with peripheral ions, but the ions of the *first* square, since they are now *contained*, are given their full charge.

Thus to a second approximation:

$$A = \frac{4 \cdot 1}{1} - \frac{4 \cdot 1}{2^{1/2}} - \frac{4 \cdot \frac{1}{2}}{4^{1/2}} + \frac{8 \cdot \frac{1}{2}}{5^{1/2}} - \frac{4 \cdot \frac{1}{4}}{8^{1/2}}$$

$$= 1{\cdot}6069$$

Subsequent approximations, by expanding the groups, leads to consecutive values of 1·6105, 1·6135, 1·6145 etc. for the Madelung constant (these values should be confirmed). The rapid convergence is very satisfactory and the values obtained at even the third or fourth level of approximation agrees well with the accepted limit of 1·61554.

We may apply this method to the sodium chloride structure by taking cubes of volume a^3, $8a^3$... (see Fig. 1.8) and ascribing fractional charges appropriately. To a first approximation we obtain

$$A = \frac{6 \cdot \frac{1}{2}}{1} - \frac{12 \cdot \frac{1}{4}}{2^{1/2}} + \frac{8 \cdot \frac{1}{8}}{3^{1/2}}$$

$$= 1{\cdot}456$$

It is left as a necessary exercise for the student to extend the calculation at least to the next stage of approximation which leads to $A = 1{\cdot}752$. The accepted limit is 1·747558.

It must be emphasised that our approach to the topic of Madelung constants has been rather sketchy; we have deliberately avoided a number of problems and further variations could be described. For further details, Tosi's paper[2] should be consulted. For convenience, however, a list of Madelung constants for some typical ionic crystals is presented in Table 1.4.

It may be of interest at this stage to use our value of the Madelung constant for sodium chloride in order to calculate the binding energy of the crystal with respect to free ions. Using the value of 2·82 Å for the nearest neighbour

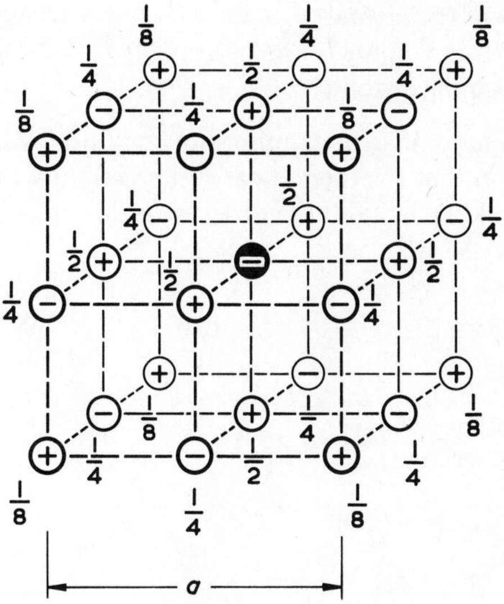

FIG. 1.8 Calculation of the Madelung constant for the sodium chloride crystal.

separation, it should be confirmed that the calculated binding energy is ~ -858 kJ mol^{-1} compared with experimental value of $-764{\cdot}5$ kJ mol^{-1}. The agreement is surprisingly good considering that short-range repulsions have not been considered. This will be our next task.

TABLE 1.4 Madelung constants

Structure	*Madelung constant*
Sodium chloride	1·74756
Caesium chloride	1·76267
Fluorite	2·51939
Zincblende	1·63805
Wurtzite	1·64132

The repulsive term

The repulsive term usually contains two adjustable parameters which may be chosen empirically by making the expression for the total energy U satisfy two conditions:

$$(i) \quad \left(\frac{\mathrm{d}U}{\mathrm{d}V}\right)_{V=V_0} = 0 \tag{1.10}$$

$$(ii) \quad \left(\frac{\mathrm{d}^2U}{\mathrm{d}V^2}\right)_{V=V_0} = \frac{1}{V_0\beta} \tag{1.11}$$

where V = volume of the crystal
V_0 = equilibrium volume of the crystal
β = crystal compressibility

Equation 1.10 is no more than an equilibrium condition and eqn. 1.11, which is merely a condition that the theoretical compressibility ($= 1/B$ where B = bulk modulus) should be equal to its observed value, may be understood as follows.

By definition

$$\frac{1}{\beta} = -V\frac{\mathrm{d}P}{\mathrm{d}V} \tag{1.12}$$

where P = pressure
Also at 0 K we may write

$$\mathrm{d}U = -P\,\mathrm{d}V \tag{1.13}$$

$$\therefore \quad \frac{\mathrm{d}P}{\mathrm{d}V} = -\frac{\mathrm{d}^2U}{\mathrm{d}V^2} \tag{1.14}$$

or

$$\frac{1}{\beta} = V\frac{\mathrm{d}^2U}{\mathrm{d}V^2} \tag{1.15}$$

Born first assumed that the short range repulsive forces between ions, for 1 mole of the crystal, could be described by a nearest neighbour interaction energy W of the form

$$W = \frac{b}{r_0{}^n} \tag{1.16}$$

where b and n are arbitrary constants. Hence the total energy of the crystal U is given by the expression

$$U = \Phi + W \tag{1.17}$$

$$= -N_0 A\frac{q^2}{r_0} + \frac{b}{r_0{}^n} \tag{1.18}$$

where for convenience we have dropped $4\pi\varepsilon_0$ from the first part of 1.18 (for those of pedantic disposition this may be done by expressing the Madelung constant in units of $4\pi\varepsilon_0$).

Next, we may express the molar volume V in terms of the ionic separation r by means of the equation

$$V = N_0\alpha r^3 \tag{1.19}$$

where α = a constant appropriate to the particular crystal structure.

Hence since

$$\frac{dU}{dV} = \frac{dU}{dr} \cdot \frac{dr}{dV},$$

we have from eqn. 1.19

$$\frac{dU}{dV} = \frac{1}{3N_0 \alpha r^2} \cdot \frac{dU}{dr} \tag{1.20}$$

and

$$\frac{d^2U}{dV^2} = \frac{1}{9N_0^2\alpha^2 r^2} \cdot \frac{d}{dr}\left(\frac{1}{r^2} \cdot \frac{dU}{dr}\right) \tag{1.21}$$

Hence from eqns 1.18, 1.20, and 1.21 and by incorporating the conditions of eqns 1.10 and 1.11 we obtain

$$\left(\frac{dU}{dV}\right)_{V=V_0} = N_0 A q^2 - \frac{nb}{r_0^{n-1}} = 0 \tag{1.22}$$

and

$$\left(\frac{d^2U}{dV^2}\right)_{V=V_0} = \frac{1}{N_0 \alpha r_0^3 \beta} = \frac{1}{9N_0^2\alpha^2}\left[\frac{n(n+3)b}{r_0^{(n+6)}} - \frac{4NAq^2}{r_0^7}\right] \tag{1.23}$$

Hence:

$$b = \frac{N_0 A q^2 r_0^{n-1}}{n} \tag{1.24}$$

$$n = \frac{1 + 9\alpha r_0^4}{Aq^2\beta} \tag{1.25}$$

and

$$U = -\frac{N_0 A q^2}{r_0}\left(1 - \frac{1}{n}\right) \tag{1.26}$$

From measured values of the compressibility, a value of n may be calculated by use of eqn. 1.25. For example, in the case of sodium chloride $n = 9.1$ (indeed except for the crystals containing very light ions a value of $n \simeq 9$ is common).

The success of simple calculations of the type described above may be seen by inspection of Table 1.5 and more recent developments have not improved significantly the agreement between calculation and experiment. Subsequent to the early calculations of Born *et al.* and the birth of quantum mechanics, however, it was recognised that the b/r^n repulsion potential had no theoretical

TABLE 1.5 Binding energies (kJ mol^{-1}) of some alkali metal halides using the b/r^n* and $a\exp(-r/\rho)$‡ repulsion potentials

Salt	$-U$ (calc.)*	$-U$ (calc.)‡	$-U$ (exp)
NaF	900·2	901·0	897·7
NaCl	755·3	747·8	764·5
NaBr	718·9	708·4	727·7
NaI	673·3	655·7	683·3
KF	797·2	791·8	794·7
KCl	688·3	676·6	694·2
KBr	660·7	646·9	663·6
KI	623·9	605·5	627·6

justification (although clearly a good approximation at short range), so it was replaced by the exponential term

$$W = a\exp\left(-\frac{r_0}{\rho}\right) \tag{1.27}$$

where a and ρ are constants. A treatment similar to that already described is clearly possible whereby these constants may be expressed in terms of β and r_0 and it is left to the reader to check the following equations:

$$a = \frac{\rho q^2 A\exp(r_0/\rho)}{{r_0}^2} \tag{1.28}$$

$$\frac{r_0}{\rho} = 2 + \frac{9\alpha {r_0}^2}{Aq^2\beta} \tag{1.29}$$

$$U = \frac{AN_0 q^2}{r_0}\left(1 - \frac{\rho}{r_0}\right) \tag{1.30}$$

The efficacy of this form of the repulsion potential is also demonstrated in Table 1.5, although agreement with experiment appears less impressive!

Further refinements

Although we shall not deal in detail with other contributions to the binding energy, it would be improper not to mention them. Van der Waals forces (which are responsible for the binding in the crystalline rare gases and may be visualised semi-classically as induced atom dipole-dipole interactions) may be described by a term proportional to $1/r^6$. Actually the $1/r^6$ term is only the first term of a series describing multipole interactions and the next term which contains $1/r^8$ (dipole-quadrupole) can contribute significantly. Fortunately the quadrupole-quadrupole term can be neglected. Finally we may mention zero point energy. Although we may not have stated it explicitly, it is obvious that our calculations are appropriate to 0 K whereby classically the kinetic energy of the ion could be neglected. According to quantum mechanics, however, even at 0 K an oscillator of frequency ω has an energy of $\hbar\omega/2$. In a solid containing $2N$ ions, there are $6N$ oscillators so that an addition to the repulsion term must be included. The influence of these terms may be ascertained from tables in Kittel and in Seitz.

References

(1) L. Pauling, *Nature of the Chemical Bond*, 3rd edition (Oxford University Press, 1959).
(2) M. P. Tosi, *Solid State Physics* **16**, 1 (1964).
(3) P. P. Ewald, *Ann. Physik* **64**, 253 (1921).

2

Crystal Symmetry

2.1 Point groups and their representation

Symmetry elements and operations

Although most people have some intuitive feeling about symmetry (for example, that one object is more symmetrical than another) our first task must be to develop a language which describes *quantitatively* the external and internal symmetry properties of crystals. By these terms we mean respectively those symmetry properties which are the consequence of (i) the observation of superficial crystalline features such as the characteristic angles between faces and (ii) information about the positions of atoms in the crystal, obtained from diffraction studies. We will approach this problem in several stages and start by considering external symmetry. Since any up-to-date course in physical science includes at least some discussion of point group symmetry, we need not dwell for too long on the elementary aspects of this subject. We will first remind ourselves of a few definitions and then explore some less familiar paths.

We begin by a consideration of symmetry elements and operations. We are naturally led in this direction by asking the important general question: why do we consider one object to be of high symmetry while another to be of low symmetry? The difference naturally lies in the difference in the number of *elements of symmetry* (such as planes through which the object may be reflected, or axes around which it may be rotated to an equivalent and thereby indistinguishable configuration). The processes of carrying out these reflexions, rotations etc. are known as *symmetry operations.* The terms symmetry element and symmetry operation must not be used synonymously.

Since use will be made entirely of the International (or Hermann-Maugin) symbolism, it may be useful to set out the symbols for symmetry elements and allowed combinations of symmetry elements in this system (see Table 2.1). For those more familiar with the Schoenflies system, these symbols may appear unattractive, but with a little practice, no difficulties should arise. In fact very few comments on Table 2.1 are required other than on the symmetry elements n and $\bar{n}$. The order of the rotation axis n is determined by the associated symmetry operation $2\pi/n$. Thus forgetting the trivial $n = 1$, all values up to seven are known for molecules, whereas crystal systems are described by $n = 2, 3, 4,$ and 6. There is a simple reason, left for the reader to

TABLE 2.1 Symmetry elements

Description of symmetry elements	*Symbol*
Rotation axis	n
Inversion axis	$\bar{n}$
Mirror plane	m
Rotation axis with mirror plane normal	n/m
Rotation axis with mirror plane parallel	nm
Rotation axis with diad axis perpendicular	$n2$
Inversion axis with diad axis perpendicular	$\bar{n}2$
Inversion axis with mirror plane parallel	$\bar{n}m$
Rotation axis with mirror plane normal and mirror planes parallel	n/mmm

consider, for the exclusion of other values in the latter case. The inversion axis $\bar{n}$ may cause some slight confusion for those used to Schoenflies symbols, since in the latter there exists the improper or rotation-reflexion axis S_n. The analogous compound operation in the International system involves instead rotation followed by inversion through a centre of symmetry. Thus an n-fold improper axis is equivalent to an $n/2$-fold inversion axis. As a further consequence of this difference of definition, the Shoenflies centre of inversion i does not require a separate International symbol since it becomes unequivocally incorporated as $\bar{1}$.

Now suppose that we have a complete list of all of the symmetry elements possessed by a molecule or crystal, how can this list be summarised to provide a convenient symbol which will indicate immediately the overall symmetry. The route is a straightforward and familiar one since *symmetry operations* satisfy the necessary criteria of constituting a mathematical group. These criteria are of sufficient importance for us to divert our attention temporarily from symmetry to some elementary definitions and theorems of group theory.

Properties of a group

A mathematical group is a set of *elements* (without any necessary physical significance) which obey certain rules which are set out below.

(1) *The product of two elements is also an element of the set.* This is fairly straightforward except for two points. First *product* in this context does not have its usual arithmetic significance as may be recognised immediately if we anticipate the outcome of this discussion and equate elements of our set to symmetry operations. Thus the product AB implies the *overall* consequence of the operation B followed by the operation A. Second the product AB is not necessarily equivalent to the product BA (i.e. $AB \neq BA$). Hence we say that A and B do not necessarily commute. There are some groups, however, in

which a commutative law does hold; these are called *Abelian* groups to which we should have cause to refer later.

(2) *The associative product rule must hold.* This is readily summarised without confusion by:

$$A(BC) = AB(C)$$

(3) *One element of the set must commute with all others to leave them unchanged.* The element with this property is usually given the symbol E, so that

$$AE = EA = A$$

(4) *Every element must have a reciprocal (or inverse) which is also an element of the set and which satisfies:*

$$AA^{-1} = A^{-1}A = E$$

Subgroups

It is convenient to demonstrate that a set of elements constitutes a group by writing the group multiplication table which is a tabular representation of the products of the elements (an element of the top *row* multiplied by a chosen element in the left hand *column*). For example:

	E	A	B	C	D	F
E	E	A	B	C	D	F
A	A	E	D	F	B	C
B	B	F	E	D	C	A
C	C	D	F	E	A	B
D	D	C	A	B	F	E
F	F	B	C	A	E	D

It is a useful exercise to confirm that the elements E, A, B, C, D, and F make up a group and this should be done by any reader not familiar with group theory. It will be observed from the group multiplication table that within this group (of order 6) there exist other groups of lower order called subgroups of order 1 (trivial E), 2, and 3. Thus it is demonstrated that (and an interesting exercise to *prove* that) the order of a subgroup is a factor of the order of the main group.

Similarity transforms

If X and Y are two elements of a group, then $X^{-1}YX$ will be equal to an element Z of the group, i.e.

$$X^{-1}YX = Z$$

This relationship is described by the statement that Z is the *similarity transform* of Y by X. We may also say that Y and Z are conjugate.

Group symmetry

Having defined a mathematical group, we can now return to our discussion of symmetry by readily demonstrating that symmetry operations satisfy the conditions of a group. We must realise, however, that this demonstration in no way constitutes a general proof, but is sufficient for our temporary needs. Let us consider as a simple example an isosceles triangle (see Fig. 2.1).

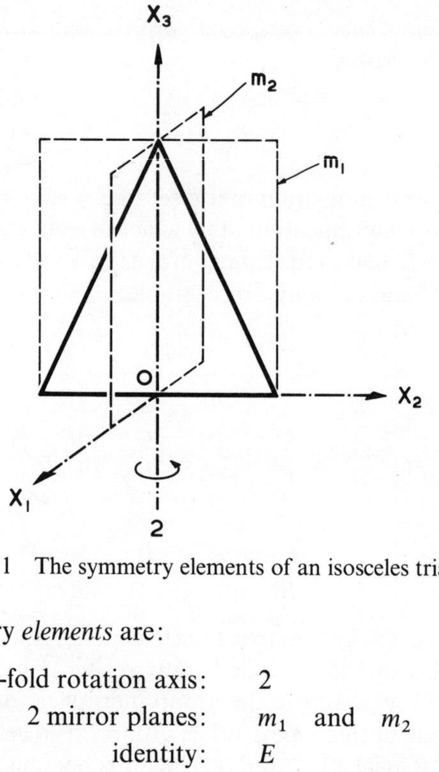

FIG. 2.1 The symmetry elements of an isosceles triangle.

The total symmetry *elements* are:

a 2-fold rotation axis:	2
2 mirror planes:	m_1 and m_2
identity:	E

Using the same symbols for the symmetry *operations*, we may look at the rules 1 to 4 as given previously.

(1) We see immediately that the combination of any two operations lead to a third: e.g.

$$2m_1 = m_2$$
$$m_1 m_2 = 2 \text{ (Do the operations commute?)}$$

(2) The associative law holds: e.g.

$$2(m_1m_2) = (2m_1)m_2$$

(3) The identity operation satisfies the condition: e.g.

$$2E = E2 = 2$$

(4) Each operation is seen to be its own inverse: e.g.

$$m_1m_1 = E$$

The example chosen for this brief treatment was particularly simple but illustrates the main points of issue. However, it is strongly suggested that the reader to whom this language is a little strange should enumerate the symmetry elements of some simple three-dimensional solids and demonstrate that the operations constitute a mathematical group.

A more satisfactory approach is to employ some elementary matrix algebra, a knowledge of which will be assumed, when we will find that a symmetry operation may be represented by a matrix and that consecutive operations may be represented by matrix products.

We define a general point (x_1, x_2, x_3) in a right-handed orthogonal coordinate system* by the vector $\boldsymbol{r}$ from the origin to that point:

$$\boldsymbol{r} = x_1\hat{\boldsymbol{x}}_1 + x_2\hat{\boldsymbol{x}}_2 + x_3\hat{\boldsymbol{x}}_3 \quad (2.2)$$

where $\hat{\boldsymbol{x}}_1$, $\hat{\boldsymbol{x}}_2$, and $\hat{\boldsymbol{x}}_3$ are the unit vectors defining the coordinate system. We consider an operation Q on this vector (fixed at the origin) to a new position $\boldsymbol{r}'$ such that

$$\boldsymbol{r}' = Q\boldsymbol{r} \quad (2.3)$$

Hence from eqn. 2.2:

$$\boldsymbol{r}' = x_1(Q\hat{\boldsymbol{x}}_1) + x_2(Q\hat{\boldsymbol{x}}_2) + x_3(Q\hat{\boldsymbol{x}}_3) \quad (2.4)$$

We may express the vectors $Q\hat{\boldsymbol{x}}_1$, $Q\hat{\boldsymbol{x}}_2$, and $Q\hat{\boldsymbol{x}}_3$ in terms of the original unit vectors by use of eqn. 2.2 leading to

$$Q\hat{\boldsymbol{x}}_1 = a_{11}\hat{\boldsymbol{x}}_1 + a_{21}\hat{\boldsymbol{x}}_2 + a_{31}\hat{\boldsymbol{x}}_3 \quad (2.5)$$

$$Q\hat{\boldsymbol{x}}_2 = a_{12}\hat{\boldsymbol{x}}_1 + a_{22}\hat{\boldsymbol{x}}_2 + a_{32}\hat{\boldsymbol{x}}_3 \quad (2.6)$$

$$Q\hat{\boldsymbol{x}}_3 = a_{13}\hat{\boldsymbol{x}}_1 + a_{23}\hat{\boldsymbol{x}}_2 + a_{33}\hat{\boldsymbol{x}}_3 \quad (2.7)$$

* In a right-hand coordinate system, the order of the axes $X_1 \rightarrow X_2 \rightarrow X_3$ is that of the pitch of a normal right-handed screw thread.

or in matrix notation:

$$Q\begin{pmatrix}\hat{\boldsymbol{x}}_1\\ \hat{\boldsymbol{x}}_2\\ \hat{\boldsymbol{x}}_3\end{pmatrix} = \begin{pmatrix}a_{11} & a_{21} & a_{31}\\ a_{12} & a_{22} & a_{32}\\ a_{13} & a_{23} & a_{33}\end{pmatrix}\begin{pmatrix}\hat{\boldsymbol{x}}_1\\ \hat{\boldsymbol{x}}_2\\ \hat{\boldsymbol{x}}_3\end{pmatrix} \tag{2.8}$$

where the matrix elements a_{ji} are the components of the vectors $Q\hat{\boldsymbol{x}}_i$ with respect to the original unit vectors.

Substitution of eqns 2.5–2.7 in eqn. 2.4 leads to

$$\begin{aligned}\boldsymbol{r}' = {} & x_1(a_{11}\hat{\boldsymbol{x}}_1 + a_{21}\hat{\boldsymbol{x}}_2 + a_{31}\hat{\boldsymbol{x}}_3) + x_2(a_{12}\hat{\boldsymbol{x}}_1 + a_{22}\hat{\boldsymbol{x}}_2 + a_{32}\hat{\boldsymbol{x}}_3)\\ & + x_3(a_{13}\hat{\boldsymbol{x}}_1 + a_{23}\hat{\boldsymbol{x}}_2 + a_{33}\hat{\boldsymbol{x}}_3)\end{aligned} \tag{2.9}$$

which is of the form

$$\boldsymbol{r}' = x'_1\hat{\boldsymbol{x}}_1 + x'_2\hat{\boldsymbol{x}}_2 + x'_3\hat{\boldsymbol{x}}_3 \tag{2.10}$$

where

$$x'_1 = a_{11}x_1 + a_{12}x_2 + a_{13}x_3 \tag{2.11}$$

$$x'_2 = a_{21}x_1 + a_{22}x_2 + a_{23}x_3 \tag{2.12}$$

$$x'_3 = a_{31}x_1 + a_{32}x_2 + a_{33}x_3 \tag{2.13}$$

Equation 2.10 indicates that (x'_1, x'_2, x'_3) are the coordinates of the transformed point which are related to the old coordinates by eqns 2.11–2.13. These latter equations may be summarised in matrix language:

$$\begin{pmatrix}x'_1\\ x'_2\\ x'_3\end{pmatrix} = \begin{pmatrix}a_{11} & a_{12} & a_{13}\\ a_{21} & a_{22} & a_{23}\\ a_{31} & a_{32} & a_{33}\end{pmatrix}\begin{pmatrix}x_1\\ x_2\\ x_3\end{pmatrix} \tag{2.14}$$

The matrix

$$(a_{ij}) = \begin{pmatrix}a_{11} & a_{12} & a_{13}\\ a_{21} & a_{22} & a_{23}\\ a_{31} & a_{32} & a_{33}\end{pmatrix} \tag{2.15}$$

uniquely defines the symmetry operation Q so may be said to *represent* that operation. Such a matrix is commonly referred to as a *transformation matrix* whose elements a_{ij} may be conveniently recognised as the *direction cosines* between new and old axes in the right-handed coordinate system taken in the order old → new (see Fig. 2.2).

Rather than by matrix formalism, there is an alternative means of summarising eqns 2.11–2.13 by use of the single equation

$$x'_i = \sum_{j=1}^{3} a_{ij}x_j \qquad (i = 1, 2, \text{or } 3) \tag{2.16}$$

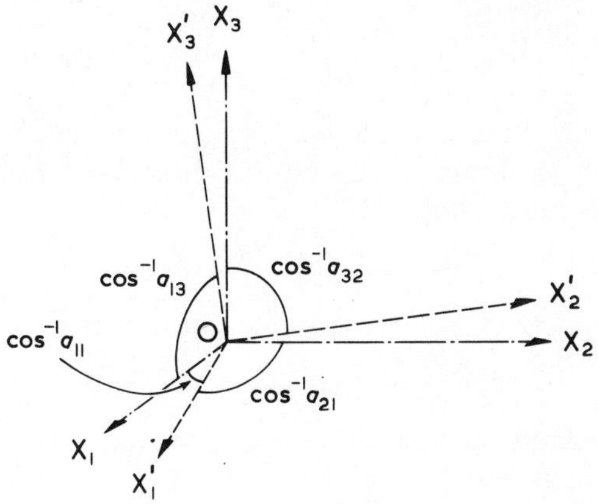

FIG. 2.2 Direction cosines.

if one adopts the so-called *summation convention* or *suffix notation.* The three equations may be obtained from 2.16 by substituting the three possible values of i. This suffix, which appears only *once* on the left and right hand sides of the equation, is called the *free suffix* since it is free to take any particular numerical value. The other suffix j, which appears twice on the right hand side of the equation, is not free since the summation sign states that we must sum over the different numerical values of this suffix. It is called the *dummy suffix* since we could use any other symbol (except i, of course) without changing the meaning of the equation. Thus

$$x_i' = \sum_{j=1}^{3} a_{ij}x_j = \sum_{k=1}^{3} a_{ik}x_k$$

It is common practice to leave out the summation sign of eqn. 2.16 to give

$$x_i' = a_{ij}x_j \qquad (i, j = 1, 2 \text{ or } 3)$$

where it is understood that if the dummy suffix appears twice in the same term, summation over the appropriate values of that suffix is automatically implied. In this book we shall use this summation convention and write

$$x_i' = a_{ij}x_j \tag{2.17}$$

and it will always be assumed that the suffices take the values 1, 2, or 3.

Since each row of the general transformation matrix above contains the direction cosines of a straight line with respect to the original axes, it follows immediately that

$$a_{11}^2 + a_{12}^2 + a_{13}^2 = 1 \tag{2.18}$$

$$a_{21}^2 + a_{22}^2 + a_{23}^2 = 1 \tag{2.19}$$

and

$$a_{31}^2 + a_{32}^2 + a_{33}^2 = 1 \tag{2.20}$$

Also since the axes of the new coordinate system are mutually perpendicular, it may be readily confirmed that a further set of relationships between the direction cosines must follow:

$$a_{11}a_{21} + a_{12}a_{22} + a_{13}a_{23} = 0 \tag{2.21}$$

$$a_{21}a_{31} + a_{22}a_{32} + a_{23}a_{33} = 0 \tag{2.22}$$

$$a_{31}a_{11} + a_{32}a_{12} + a_{33}a_{13} = 0 \tag{2.23}$$

We may conveniently summarise these two sets of equations by means of the summation convention.

From 2.18–2.20:

$$a_{ik}a_{jk} = 1 \qquad (i = j) \tag{2.24}$$

and from 2.21–2.23:

$$a_{ik}a_{jk} = 0 \qquad (i \neq j) \tag{2.25}$$

Equations 2.24 and 2.25 may be combined by use of δ_{ij}, which is called the Kronecker delta and is defined by

$$\delta_{ij} = \begin{cases} 0 & \text{if } i \neq j \\ 1 & \text{if } i = j \end{cases}$$

Therefore we write

$$a_{ik}a_{jk} = \delta_{ij} \tag{2.26}$$

Relationships of the type 2.26 are known as *orthogonality relationships* and the transformation whose matrix elements satisfy orthogonality conditions are called linear orthogonal transformations.

Let us consider two consecutive orthogonal transformations of a point

$$(x_1 x_2 x_3) \xrightarrow{\text{(i)}} (x_1' x_2' x_3') \xrightarrow{\text{(ii)}} (x_1'' x_2'' x_3'')$$
$$\underbrace{\phantom{(x_1 x_2 x_3) \xrightarrow{} (x_1' x_2' x_3') \xrightarrow{} (x_1'' x_2'' x_3'')}}_{\text{(iii)}}$$

Transformations (i), (ii) and (iii) may be represented respectively by:

$$x_i' = a_{ij}x_j \tag{2.27}$$

$$x_i'' = b_{ij}x_j' \tag{2.28}$$

$$x_i'' = c_{ij}x_j \tag{2.29}$$

We leave the reader to confirm by comparison of coefficients that

$$c_{ij} = a_{ik}b_{kj} \tag{2.30}$$

(if in doubt, 2.30 may be confirmed by expansion).

Equation 2.30 will be recognised as equivalent to a statement of the product rule for two matrices, so that we may now presume that the matrix product describing two consecutive operations must be equivalent to the matrix describing the *overall* transformation. We note also that c_{ij} satisfies the orthogonality conditions.

Before leaving this general topic, we should mention a useful property of the transformation determinant, (Δ) which is no more than the determinant of the usual array of direction cosines i.e.

$$\Delta = \begin{vmatrix} a_{11} & a_{12} & a_{13} \\ a_{21} & a_{22} & a_{23} \\ a_{31} & a_{32} & a_{33} \end{vmatrix}$$

By judicious application of the orthogonality relationship and elementary determinant manipulation, the reader should show that

$$\Delta = \pm 1 \tag{2.31}$$

Equation 2.31 is a statement of the well-known observation that for all transformations from one rectangular coordinate system to another, the transformation determinant is equal to ± 1. If for no other reason it is a useful check.

After the relative complexity of the last few pages, it may seem trivial now to return to our isosceles triangle (Fig. 2.1). Nevertheless it is a worthwhile exercise to write down the transformation matrices and to confirm their relationship.

The matrices corresponding to the operations are:

$$2 \equiv \begin{pmatrix} -1 & 0 & 0 \\ 0 & -1 & 0 \\ 0 & 0 & 1 \end{pmatrix}$$

$$m_1 \equiv \begin{pmatrix} -1 & 0 & 0 \\ 0 & 1 & 0 \\ 0 & 0 & 1 \end{pmatrix}$$

$$m_2 \equiv \begin{pmatrix} 1 & 0 & 0 \\ 0 & -1 & 0 \\ 0 & 0 & 1 \end{pmatrix}$$

$$E \equiv \begin{pmatrix} 1 & 0 & 0 \\ 0 & 1 & 0 \\ 0 & 0 & 1 \end{pmatrix}$$

Also by the usual rules of matrix algebra, we may demonstrate that, for example: $2m_1 = m_2$. Other relationships should be checked.

Crystal systems and crystal classes

In the previous section we have made an effort to demonstrate that symmetry operations appropriate to some object constitute a mathematical group. This group of symmetry operations, all of which leave at least one point invariant is known as the *point group*. Although there may be a number of symmetry operations that one can carry out on some geometric figure, only a few are independent in the sense that when two or three of them are specified, the others must *necessarily* follow. The independent operations are frequently referred to as the *generating set*. To label any point group we need only to state the generating set. Any single crystal must belong to one of 32 point groups known as the *crystal classes*, each class belonging to one of 7 *crystal systems*. (See Table 2.2).

TABLE 2.2 The crystal classes and crystal systems

Rotational symmetry	*Crystal system*	*Crystal classes*
1	Triclinic	$1; \bar{1}$
2	Monoclinic	$2; m; 2/m$
Three × 2	Orthorhombic	$mm2; 222; mmm$
3 or $\bar{3}$	Trigonal	$3; \bar{3}; 3m; 32; \bar{3}m$
4 or $\bar{4}$	Tetragonal	$4; \bar{4}; 4/m; 4mm; 42m; 422; 4/mmm$
6 or $\bar{6}$	Hexagonal	$6; \bar{6}; 6/m; 6mm; \bar{6}m2; 622; 6/mmm$
Four × 3	Cubic	$23; m3; \bar{4}3m; 432; m3m$

2.2 The stereographic projection

Casual inspection of a crystal may be insufficient to assign it to a crystal class since the irregular size and shape of the faces can obscure the true symmetry which is determined by the *angular* relationships between the faces. A convenient representation which suppresses the chance variation in sizes of the crystal faces is a projection on to a sphere of all the normals to the faces from a common origin. As a simple example, we show in Fig. 2.3 the projections of the normals to the faces of a regular octahedron. The normals are, of course, the four 3-fold axes of the cubic system. The octahedron has point group symmetry *m3m*. Such a spherical projection is still inconvenient and we require some related two-dimensional representation. This is the stereographic projection whose construction we may now describe. Again a sphere is supposed to surround the crystal and from the centre of the sphere, normals are drawn through the crystal faces. The point at which the normal to each face meets the sphere is then joined by a line through the equatorial plane (which is the plane of projection) to the pole of the lower hemisphere.

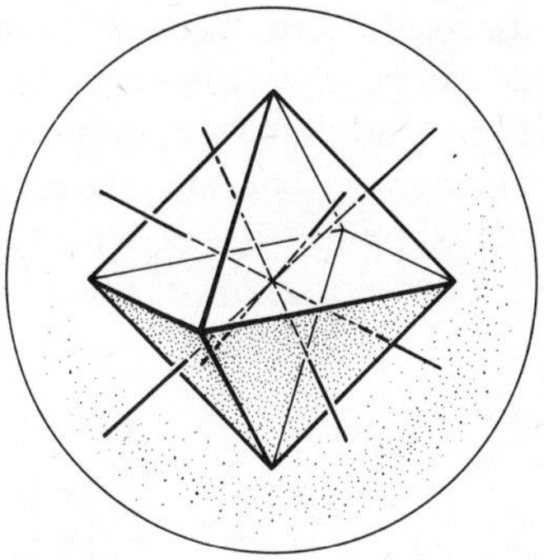

FIG. 2.3 Projection of an octahedron on to an enclosing sphere.

The points where these lines cut the equatorial plane (sometimes also called the plane of the primitive) are the stereographic poles of the corresponding crystal faces and are shown by a blocked-in circle. Where normal projection meets the lower hemisphere, projection through the equatorial plane to the upper pole is employed. To distinguish projections to the upper pole from those to the lower pole, the former are marked by open circles. The procedure is illustrated in Fig. 2.4.

Clearly, there must be certain geometrical relationships associated with the projection which we could discuss, but since they are of no direct relevance to our needs, we refer the interested reader to Phillips. What does concern us, however, is the usefulness of the stereographic projection in representing the symmetry of a crystal; we may even say in representing directly the point group itself. Since the relative positions of the projections on the sphere determine the feasible symmetry operations, thus the poles of the stereogram equally represent the symmetry elements of the crystal, i.e. its point group. We may therefore add to the stereogram representations of the symmetry elements relating the poles according to a system of shorthand, the details of

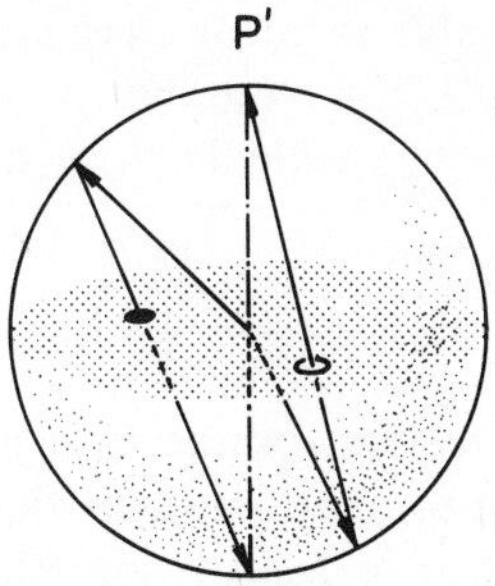

FIG. 2.4 The stereographic projection.

which will not concern us. Stereograms of the 32 point groups showing symmetry elements and poles may be found in any text-book on crystallography.

The usefulness of this representation will become apparent on construction of stereograms, a recommended exercise; this is not a difficult exercise, although it may get a little complicated, say, with cubic groups. As a simple example let us construct the stereogram of one common point group 32.

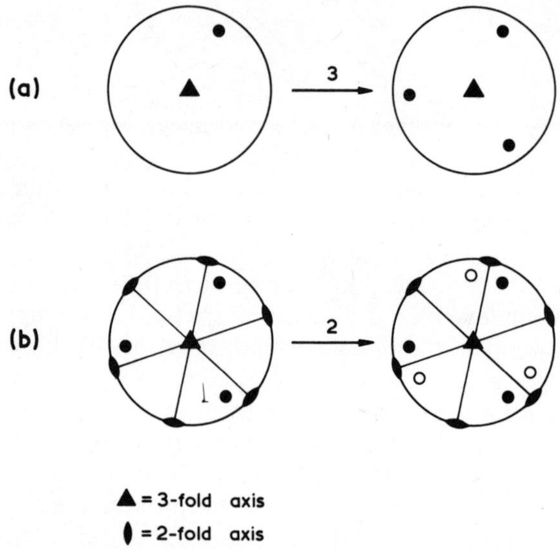

FIG. 2.5 Stereogram of point group 32.

We first choose a general pole and operate on it according to the group. The 3-fold axis is taken along PP' so that the general pole is transformed to trigonally equivalent positions [see Fig. 2.5(a)]. The operation of the three 2-fold rotations (in the primitive plane perpendicular to 3) then transfers the poles according to Fig. 2.5(b). Note that the transformed poles are open circles since each operation of 2 will effect transfer to the other hemisphere.

2.3 Symmetry and the physical properties of crystals

Having established some feeling for the external symmetry of crystals it should be possible to begin to consider in an elementary way a topic which has, at least, the appearance of having something to do with solid state physics: the effect of crystal orientation on physical properties. In order to discuss this topic quantitatively we require the application of tensor algebra.

Tensors

Measurements made on a solid are related by its physical properties which may be *scalars*, *vectors* or *tensors*. For example, the scalar quantities mass (m) and volume (v) are related by the scalar property density (ρ), but when one of the measured quantities has direction, the property must be a vector, as in the pyroelectric effect (the temperature dependence of the polarisation of a crystal). Extending the argument, it is apparent that if both the measured quantities are direction dependent, the corresponding property must have a more complex form: in fact that of a second or higher rank tensor.

Let us begin with a specific example: the variation of induced dipole moment with crystal direction in an applied electric field. We will assume that for some appropriately chosen direction, the induced dipole per unit volume (or polarisation) $\boldsymbol{P}$ is proportional to the applied field $\mathscr{E}$. Not being concerned about units, we will label the constant of proportionality α, the polarisability. Since we observe that both $\boldsymbol{P}$ and $\mathscr{E}$ are vectors, we require information about their orientations with respect to each other and with respect to that of the crystal. This is the context in which tensor alebra is of supreme utility. Let us suppose by way of introduction that by a fortunate choice of crystal orientation, an electric field $\mathscr{E}_1$, in the X_1 direction produces a polarisation P_1 in the same direction. If we now apply an electric field E_2 in the X_2 direction of equal magnitude to that of $\mathscr{E}_1$, the polarization P_2 will not be of magnitude equal to that of P_1 since in general the crystal will not be isotropic. This implies, therefore, that application of a field $\mathscr{E}$ in the X_1X_2 plane will produce a polarisation $\boldsymbol{P}$ of orientation different from that of the field. Thus if the choice of axes had been less fortunate, the polarisation produced by $\mathscr{E}_1$ would have had also components in the X_2 and X_3 directions.

With these considerations in mind, therefore, let us consider the general case of an arbitrary orientation of the crystal with respect to the coordinate axes. We let the components of the field $\mathscr{E}$ along the X_1X_2 and X_3 directions be E_1, E_2 and E_3 respectively. Thus the component of the field $\mathscr{E}_1$, will produce a polarisation with components in the X_1X_2 and X_3 directions as indicated by 2.32

$$P'_1 = \alpha_{11}\mathscr{E}_1;\ P'_2 = \alpha_{21}\mathscr{E}_1;\ P'_3 = \alpha_{31}\mathscr{E}_1 \tag{2.32}$$

(the choice of suffices of α is an obvious one: the first refers to the polarisation component and the second to the electric field component). Similarly the polarisation components which result from the other two components of the electric field E_2 and E_3 will be respectively:

$$P''_1 = \alpha_{12}\mathscr{E}_2;\ P''_2 = \alpha_{22}\mathscr{E}_2;\ P''_3 = \alpha_{32}\mathscr{E}_2 \tag{2.33}$$

and

$$P'''_1 = \alpha_{13}\mathscr{E}_3;\ P'''_2 = \alpha_{23}\mathscr{E}_3;\ P'''_3 = \alpha_{33}\mathscr{E}_3 \tag{2.34}$$

Now the total polarisation $\boldsymbol{P}$ will have components P_1, P_2 and P_3 where

$$P_1 = P'_1 + P''_1 + P'''_1 \text{ etc}$$

Therefore

$$P_1 = \alpha_{11}\mathscr{E}_1 + \alpha_{12}\mathscr{E}_2 + \alpha_{13}\mathscr{E}_3 \tag{2.35}$$

$$P_2 = \alpha_{21}\mathscr{E}_1 + \alpha_{22}\mathscr{E}_2 + \alpha_{23}\mathscr{E}_3 \tag{2.36}$$

$$P_3 = \alpha_{31}\mathscr{E}_1 + \alpha_{32}\mathscr{E}_2 + \alpha_{33}\mathscr{E}_3 \tag{2.37}$$

The dielectric behaviour of the crystal is thus described completely by the nine coefficients α_{ij} commonly written as at array in *square* brackets (in order to differentiate it from a matrix)

$$\begin{bmatrix} \alpha_{11} & \alpha_{12} & \alpha_{13} \\ \alpha_{21} & \alpha_{22} & \alpha_{23} \\ \alpha_{31} & \alpha_{32} & \alpha_{33} \end{bmatrix}$$

and referred to as a *tensor of the second rank* (in this specific case, we refer to it as the polarisability tensor). α_{ij} are usually referred to as the *components* of the tensor; the first suffix refers to the row of the tensor and the second to the column.

The logic of referring to $[\alpha_{ij}]$ as a second rank tensor becomes apparent by comparison with three other quantities commonplace in physics.

(*i*) A *scalar* (or a tensor of *zero* rank) is specified by a single component without reference to any axial system.

(*ii*) A *vector* (or a tensor of *first* rank) is specified by three components each of which is related to *one* axis of reference.

(*iii*) A tensor of the *second* rank is specified by nine components each of which is related to *two* axes of reference.

(*iv*) A tensor of the *third* rank is specified by twenty-seven components each of which is related to *three* axes of reference.

Transformation of the components of a second rank tensor

Returning to our example of the polarisability tensor, we may enquire about the consequences of changing our original coordinate system from $X_1X_2X_3$ to a new one $X'_1X'_2X'_3$ by rotation about the origin. The components ($\mathscr{E}'_1$, $\mathscr{E}'_2$ and $\mathscr{E}'_3$) of our electric field $\mathscr{E}$ will be quite different as will the components (P'_1, P'_2 and P'_3) of the polarisability. But since we have done nothing to the system, except change the frame of reference, it must follow that the components of the polarisability tensor must change in some definable way according to the chosen coordinate system. Alternatively, we may think about the effect of changing the orientation of the crystal in the electric field. Again the components of the polarisability tensor must change in some way which is important to understand. In summary, therefore, knowing the tensor components for one orientation, we need to be able to calculate them for any other.

We recall from our previous discussion in this chapter that the components of a vector transform according to

$$r'_i = a_{ij} r_j \tag{2.38}$$

where r'_i are the components of the transformed vector $\boldsymbol{r}'$ and r_j are the components of the original vector $\boldsymbol{r}$.

We may note also (and it is easy to check) that the reverse transformation $\boldsymbol{r}' \rightarrow \boldsymbol{r}$ may be represented by the equation

$$r_i = a_{ji} r'_j \tag{2.39}$$

(note the change in the order of the suffices in 2.39, which is a self-evident consequence of the shift from rows into columns in the transformation matrix).

If we choose to write a general second rank tensor equation in the form

$$p_i = T_{ij} q_j \tag{2.40}$$

where T_{ij} are the components of the property tensor linking vectors $\boldsymbol{p}$ and $\boldsymbol{q}$, the approach to our problem is a fairly obvious one: we need only investigate the consequences of a change of axial system from $X_1X_2X_3$ to $X'_1X'_2X'_3$ so that the vector components p_i and q_i change respectively to p'_i and q'_i. This should then relate the tensor components T_{ij} and T'_{ij}. We may carry through this argument by relating sequentially the vector components $p'_i \leftarrow p_i \leftarrow q_i \leftarrow q'_i$. Clearly we will finally obtain the required tensor relationship between the components p'_i and q'_i. From 2.38, 2.39 and 2.40:

$$p'_i = a_{ik} p_k \tag{2.41}$$

$$p_k = T_{kl} q_l \tag{2.42}$$

$$q_l = a_{jl} q'_j \tag{2.43}$$

From 2.40–2.43, we see immediately that

$$p'_i = T'_{ij} q'_j \tag{2.44}$$

$$= a_{ik} T_{kl} a_{jl} q'_j \tag{2.45}$$

Comparison of eqns 2.44 and 2.45 leads to the required relationship

$$T'_{ij} = a_{ik} a_{jl} T_{kl} \tag{2.46}$$

the very important transformation rule for second rank tensors.

The effect of crystal symmetry

As a welcome diversion from definitions and mathematics, and before going on to discuss the representation quadric, we are now in a position to consider the question of the influence of the symmetry of a crystal on its physical properties: or more precisely on the form of its second rank property tensors. But first we must say something about the symmetry of second rank tensors. All of the tensors with which we will be dealing are *symmetrical.* That is they are of the form

$$\begin{bmatrix} T_{11} & T_{12} & T_{13} \\ T_{12} & T_{22} & T_{23} \\ T_{13} & T_{23} & T_{33} \end{bmatrix}$$

and thus contain only *six* independent components. We will not attempt to prove this although the implications are conceptually acceptable (we leave the reader to consider the simple consequences).

The guiding principle which allows us to consider in a very simple way the effect of crystal symmetry is known as *Neumann's Principle*, which states that the symmetry elements of any physical property of a crystal must include those of the point group of the crystal. We may now investigate the consequences of this principle by considering examples of the seven crystal systems. The simplifying feature that we need consider only the crystal systems rather than crystals of individual point groups is a direct consequence of our limiting our discussion to *symmetrical* second rank tensors. The simplification would be unacceptable otherwise.

We deal specifically with two crystal systems.

(1) *The triclinic system.* Since there are no elements of symmetry other than 1, there are no restrictions on the symmetry of the property tensor which is thus:

$$\begin{bmatrix} T_{11} & T_{12} & T_{13} \\ T_{12} & T_{22} & T_{23} \\ T_{13} & T_{23} & T_{33} \end{bmatrix}$$

(2) *The monoclinic system.* This is characterised by the element 2, taken as the axis OX_2. The transformation matrix for the operation is

$$(a_{ij}) = \begin{pmatrix} -1 & 0 & 0 \\ 0 & 1 & 0 \\ 0 & 0 & -1 \end{pmatrix}$$

By use of eqn. 2.46 we may write the components of the transformed tensor

(*i*) $T'_{11} = a_{11}a_{11}T_{11}$ (all other products containing $a_{1k}a_{1l} = 0$)
$= T_{11}$

(*ii*) $T'_{12} = a_{11}a_{22}T_{12}$ (all other products containing $a_{1k}a_{2l} = 0$)
$= -T_{12}$

(*iii*) $T'_{13} = a_{11}a_{33}T_{13}$ (all other products containing $a_{1k}a_{3l} = 0$)
$= T_{13}$

(*iv*) $T'_{22} = a_{22}a_{22}T_{22}$ (all other products containing $a_{2k}a_{2l} = 0$)
$= T_{22}$

(*v*) $T'_{23} = a_{22}a_{33}T_{23}$ (all other products containing $a_{2k}a_{3l} = 0$)
$= -T_{23}$

(*vi*) $T'_{33} = a_{33}a_{33}T_{33}$ (all other products containing $a_{3k}a_{3l} = 0$)
$= T_{33}$

Inspection of the transformed tensor components (*i*)–(*vi*) will reveal that all are unchanged by the symmetry operation 2 except (*ii*) and (*v*) which change sign. This is only allowed by Neumann's Principle if both $T'_{12} = T_{12} = 0$ and $T'_{23} = T_{23} = 0$. The property tensor appropriate to monoclinic crystals is then of the form

$$\begin{bmatrix} T_{11} & 0 & T_{13} \\ 0 & T_{22} & 0 \\ T_{13} & 0 & T_{33} \end{bmatrix}$$

in which the number of independent components is reduced to four.

We have dealt *in detail* with one system knowing that expansion of coefficient products may sometimes cause confusion; but it would be somewhat repetitive to evaluate similarly the components of the tensor of all the other crystal systems. The unfamiliar reader is strongly advised to carry out several summations independently. For example, the trigonal, hexagonal, and tetragonal systems are tackled in a fashion identical with that used in the monoclinic case. Also it should be checked that the operation 2 with reference to the OX_1 and OX_3 axes leads to the tensors

$$\begin{bmatrix} T_{11} & 0 & 0 \\ 0 & T_{22} & T_{23} \\ 0 & T_{23} & T_{33} \end{bmatrix} \quad \text{and} \quad \begin{bmatrix} T_{11} & T_{12} & 0 \\ T_{12} & T_{22} & 0 \\ 0 & 0 & T_{33} \end{bmatrix} \quad \text{respectively}$$

so that the tensor appropriate to orthorhombic crystals must be

$$\begin{bmatrix} T_{11} & 0 & 0 \\ 0 & T_{22} & 0 \\ 0 & 0 & T_{33} \end{bmatrix}$$

by Neumann's Principle.

Although it might have been more apposite to deal generally in the previous section with the topic of the effect of an operation on the crystal which is *not* a symmetry operation, it is left until this time for two reasons. First by taking one of the crystal systems (other than triclinic) the number of tensor components is conveniently reduced and second it is preferable to have had first some experience of simple product summations, before tackling this slightly more complicated problem. The serious student is urged to follow through at least one such calculation; for example, he or she could check that a rotation 3 about OX_3 in the orthorhombic system leads to the following tensor

$$\begin{bmatrix} \frac{1}{4}(T_{11} + 3T_{22}) & \frac{\sqrt{3}}{4}(T_{22} - T_{11}) & 0 \\ \frac{\sqrt{3}}{4}(T_{22} - T_{11}) & \frac{1}{4}(3T_{11} + T_{22}) & 0 \\ 0 & 0 & T_{33} \end{bmatrix}$$

The representation quadric (we follow here an approach very close to that of Nye[1])

Let us consider the equation

$$S_{ij}x_i x_j = 1 \tag{2.47}$$

Assuming the symmetry $S_{ij} = S_{ji}$, eqn. 2.47 may be expanded to give

$$S_{11}x_1^2 + S_{22}x_2^2 + S_{33}x_3^2 + 2S_{12}x_1x_2 + 2S_{13}x_1x_3 + 2S_{23}x_2x_3 = 1 \tag{2.48}$$

which is the equation of a quadric (in general an ellipsoid or hyperboloid).

Equation 2.47 may be transformed to new axes X_i' by means of the equations

$$x_i = a_{ki}x_k' \tag{2.49}$$

and

$$x_j = a_{lj}x_l' \tag{2.50}$$

thus obtaining

$$S_{ij}a_{ki}a_{lj}x'_k x'_l = 1 \tag{2.51}$$

which may be written in the form

$$S'_{kl}x'_k x'_l = 1 \tag{2.52}$$

where

$$S'_{kl} = a_{ki}a_{lj}S_{ij} \tag{2.53}$$

If eqn. 2.53 is compared with the second rank tensor transformation law (2.46) it is seen to be identical in form; so it follows that the *coefficients* of the general quadric transform like the components of a symmetrical tensor of rank two. Thus if we wish to examine how the components of a tensor transform, we may alternatively consider the transformation of the corresponding quadric.

When the general quadric is referred to its principal axes, its equation reduces to the simple form

$$S_{11}x_1{}^2 + S_{22}x_2{}^2 + S_{33}x_3{}^2 = 1 \tag{2.54}$$

Likewise the symmetrical second rank tensor is reduced to the form

$$\begin{bmatrix} T_{11} & 0 & 0 \\ 0 & T_{22} & 0 \\ 0 & 0 & T_{33} \end{bmatrix}$$

when referred to its principal axes. T_{11}, T_{22} and T_{33} (often written T_1, T_2 and T_3 respectively) are referred to as the *principal components* of the tensor.

Equation 2.54 may be written

$$\frac{x_1}{(S_{11}^{-1/2})^2} + \frac{x_2}{(S_{22}^{-1/2})^2} + \frac{x_3}{(S_{33}^{-1/2})^2} = 1 \tag{2.55}$$

so that it is apparent that the semi-axes of the representation quadric are of magnitude $S_{11}^{-1/2}$, $S_{22}^{-1/2}$ and $S_{33}^{-1/2}$.

Three surfaces may be identified (see Fig. 2.6):

(*i*) if the coefficients S_{11}, S_{22} and S_{33} are all positive the surface is an ellipsoid.

(*ii*) if two coefficients are positive and one is negative, the surface is a hyperboloid of one sheet.

(*iii*) if one coefficient is positive and two are negative, the surface is a hyperboloid of two sheets.

(The fourth combination, of course, produces an imaginary solid.)

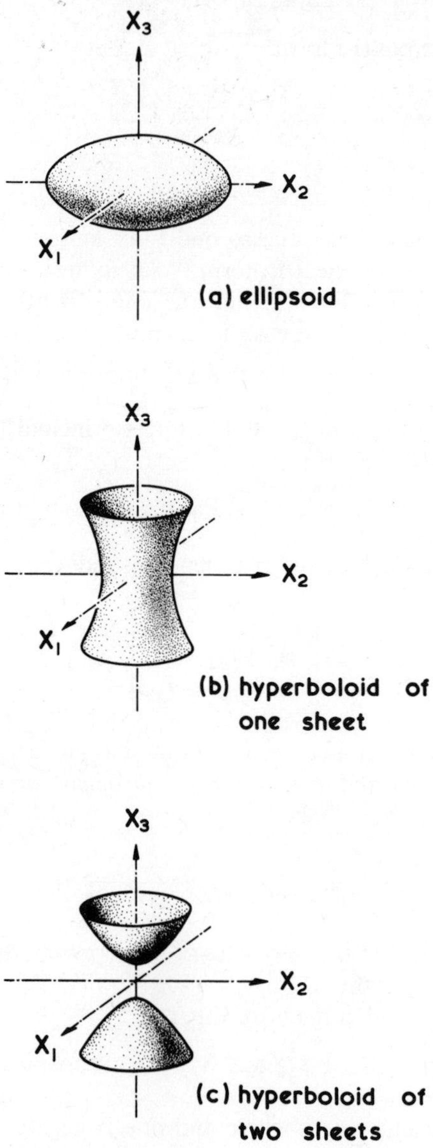

FIG. 2.6 Representation quadrics.

We may return to the question of the relationship between crystal symmetry and property tensors, since we are now equipped with a particularly simple geometric interpretation of the transformation properties of a symmetrical second rank tensor: we must ensure that the symmetry of the crystal system is shown by that of the representation quadric. We examine some crystal systems and leave the reader to make good any omissions of significance.

(*i*) *Cubic system.* The representation quadric can only conform with the requirement of having four 3-fold axes by being a sphere so that the property tensor takes the form

$$\begin{bmatrix} T_{11} & 0 & 0 \\ 0 & T_{11} & 0 \\ 0 & 0 & T_{11} \end{bmatrix}$$

Incidentally, this result is of considerable importance since it implies that where second rank tensor properties are concerned, cubic crystals are isotropic.

(*ii*) *Uniaxial systems* (*tetragonal, hexagonal and trigonal*). The only way the representation quadric can conform with the requirement of containing a 3-, 4-, or 6-fold axis is for the axis to become the principal axis (X_3) and for the quadric to become a surface of revolution about it. Two lengths, those of the major and minor axes are sufficient to define the quadric. The property tensor is thus

$$\begin{bmatrix} T_{11} & 0 & 0 \\ 0 & T_{11} & 0 \\ 0 & 0 & T_{33} \end{bmatrix}$$

The magnetic permeability tensor

Before leaving the question of second rank tensors, let us take an example of the formulation of an anisotropic property in tensor language by considering the simple (not ferro-) magnetic properties of crystals, a topic of which most chemists will have had some experience. We may express the relationship between the magnetic field strength $\boldsymbol{H}$, the magnetic induction $\boldsymbol{B}$ and the magnetisation $\boldsymbol{M}$ by the equation

$$\boldsymbol{B} = \mu_0(\boldsymbol{H} + \boldsymbol{M}) \tag{2.56}$$

where μ_0 is the permeability of a vacuum.

For many materials, $\boldsymbol{M}$ and $\boldsymbol{H}$ may be related by the tensor equation

$$M_i = \mu_0 \mathscr{K}_{ij} H_j \tag{2.57}$$

where $\mathscr{K}$ is the magnetic susceptibility tensor.

Further, since we may write 2.56 in the component form

$$B_i = \mu_0 (H_i + M_i) \tag{2.58}$$

it follows that

$$B_i = \mu_0 (H_i + \mathscr{K}_{ij} H_j) \tag{2.59}$$

The form of eqn. 2.59 may be further simplified by making use of the relationship of the form

$$\delta_{ij} r_i = r_j \tag{2.60}$$

so that

$$B_i = \mu_0 (\delta_{ij} + \mathscr{K}_{ij}) H_j \tag{2.61}$$

Hence

$$B_i = \mu_{ij} H_j \tag{2.62}$$

where the permeability tensor

$$\mu_{ij} = \mu_0 (\delta_{ij} + \mathscr{K}_{ij}) \tag{2.63}$$

In expanded form we may write

$$\begin{bmatrix} \mu_{11} & \mu_{12} & \mu_{13} \\ \mu_{21} & \mu_{22} & \mu_{23} \\ \mu_{31} & \mu_{32} & \mu_{33} \end{bmatrix} = \mu_0 \begin{bmatrix} 1 + \mathscr{K}_{11} & \mathscr{K}_{12} & \mathscr{K}_{13} \\ \mathscr{K}_{21} & 1 + \mathscr{K}_{22} & \mathscr{K}_{23} \\ \mathscr{K}_{31} & \mathscr{K}_{32} & 1 + \mathscr{K}_{33} \end{bmatrix}$$

The reader is advised to investigate other simple second rank property tensors as can be found in many text-books on general physics. Further, for those who *enjoy* manipulating matrices and tensors, tackling a third rank tensor (piezo-electric moduli, for example)[1] might prove attractive. The representation of a third rank tensor, which contains in three dimensions, 27 components, is an interesting problem in its own right.

2.4 The internal symmetry of crystals

Hitherto in our discussion of crystal symmetry and crystal properties, we have made no mention of the internal arrangement of the constitutent atoms. This will be our next task.

As mentioned in the Preface, it had been long suspected before 1912 that the characteristic external symmetry of crystals resulted from a regular arrangement of the basic "building blocks" (unspecified). The periodicity inherent in the crystalline state was finally demonstrated by von Laue[2] in 1912.

Although *real* crystals contain numerous defects (to be discussed in Chapter 7), an *ideal* crystal is characterised by a regular three-dimensional arrangement of identical structural units. These elementary units are parallelepipeds which, in the simple crystal of a metal contain one atom, or in more complex crystals contain many atoms or groups of atoms. The regular array of the parallelepipeds, each of which may be represented by a *point* in space, leads to a periodic three-dimensional display of points known as the *space lattice*. It is worth emphasising that the *space lattice has no physical reality*. It provides only information about the periodicity of the lattice, but when details of the atomic environment of each lattice point (its *basis*) are known, we obtain the reality of the *crystal structure*.

Translation vectors and unit cells

The lattice points, as described above, may be defined in an ideal crystal in terms of three fundamental translation vectors $\boldsymbol{a}_1$, $\boldsymbol{a}_2$ and $\boldsymbol{a}_3$ such that the atomic environment at all points defined by the terminal of vector $\boldsymbol{R}$ are identical. Thus

$$\boldsymbol{R} = n_1\boldsymbol{a}_1 + n_2\boldsymbol{a}_2 + n_3\boldsymbol{a}_3 \tag{2.64}$$

where n_1, n_2 and n_3 are integers.

If *every* environmentally equivalent point of the lattice is described by the vector $\boldsymbol{R}$, then $\boldsymbol{a}_1$, $\boldsymbol{a}_2$ and $\boldsymbol{a}_3$ are said to be the *primitive* translation vectors and also the lattice is said to be primitive. Alternatively, the translation vectors may be chosen not to be primitive but in order to emphasise some three-dimensional unit of convenient symmetry such as a cube.

The parallelepiped defined by the primitive translation vectors is known as the *primitive cell* (note it contains one lattice point only) which must be the *simplest unit cell*. In general, a unit cell, if translated by $\boldsymbol{R}$ will fill the entire space of the crystal, so we may think of the primitive cell as a unit cell of minimum volume. As indicated above, the unit cell may be chosen differently to conform with a convenient geometry; in this case it contains more than one lattice point. Should any confusion surround these definitions, examples of primitive and non-primitive unit cells of a two-dimensional lattice may help to clarify the situation (see Fig. 2.7). The cells P are examples of primitive cells (each containing one lattice point since each of the latter at the corners of each cell is shared by four cells). The cells N.P. are examples of a non-primitive cell containing two lattice points. Although it is of no

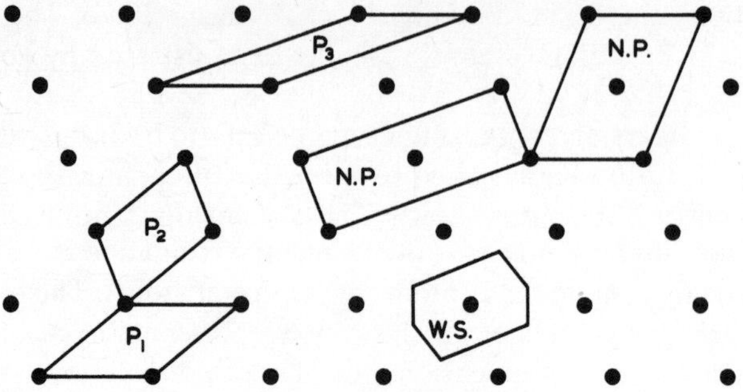

FIG. 2.7 Two-dimensional unit cells.

immediate consequence we may note a different way of dividing our two-dimensional lattice: by lines perpendicularly bisecting the lines joining adjacent lattice points (cell W.S.). These primitive unit cells are referred to as *Wigner–Seitz cells* and will be discussed later. It should be noted that Wigner–Seitz cells offer the practical and aesthetic value of retaining the symmetry of the lattice.

Bravais lattices

Superficial inspection of eqn. 2.64 will suggest that it might be possible to construct an infinite variety of lattices, but it may be assumed (without proof) that in two dimensions only *five* and in three dimensions only *fourteen* fundamentally different lattices are possible. These are the *Bravais lattices* (Figs 2.8 and 2.9). The route to a proof lies in group theory but we will not pursue the question of translational symmetry at present.

Referring to Fig. 2.9 some comments seem appropriate. Of the cubic cells only one (P) is primitive, the body-centred (I) and the face-centred (F) containing two and four lattice points per unit cell respectively. Here we note that non-primitive unit cells have been chosen to emphasise the cubic symmetry. We will consider the corresponding primitive cells at a later stage. It will be noted that only two tetragonal Bravais cells are possible; a casual reaction might be that other types of centering would still preserve the necessary 4-fold symmetry. However, as can be seen in Fig. 2.10(a), inclusion of lattice points on the 4-fold axis do not result in a new type of cell, but only a primitive cell of different orientation. Again centering the prism faces as allowed by the 4-fold symmetry proves not to constitute a true cell since we see in Fig. 2.10(b) that some of the centred lattice points have different environments. Finally [Figure 2.10(c)] an F-centred cell only produces an I-tetragonal Bravais cell of different orientation. In the orthorhombic

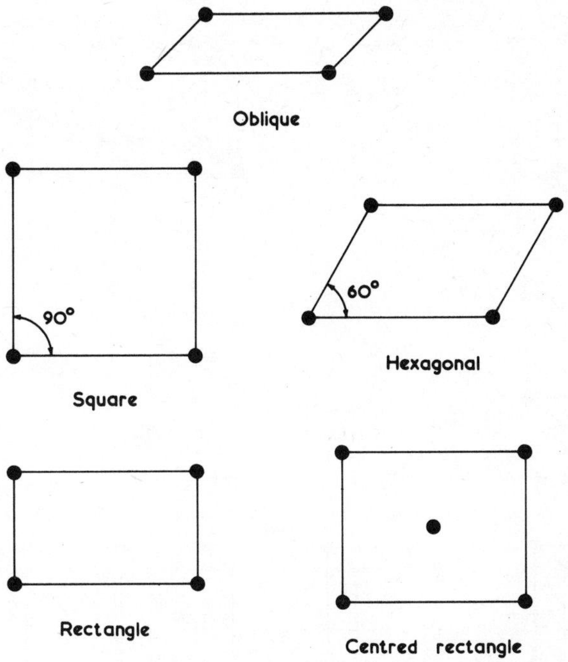

FIG. 2.8 Bravais lattices in two dimensions.

system, the simplest arrangement is the primitive cell; centering of any opposite faces preserves the 3 × 2-fold symmetry, but since all three would be equivalent, only the C-centred cell is chosen by convention. I and F centering is allowed and produces new lattice types. In the monoclinic system, besides the primitive cell, a C-lattice is allowed but centering of the B faces only produces a different primitive cell; A-face centering only produces the cell equivalent to that produced by C-centering.

The reader is advised to confirm these points and if further discussion of Bravais cells is thought appropriate, more details can be found in Phillips.

Space groups

The fourteen three-dimensional Bravais lattices all conform with the rotational symmetry of the seven crystal systems. We have indicated previously, however, that the internal crystal symmetry may be lower and clearly this lowering of symmetry must result from the disposition of atoms around the lattice points i.e. the basis. In other words, we require a new group, called the *space group* which contains the combined operations of the appropriate

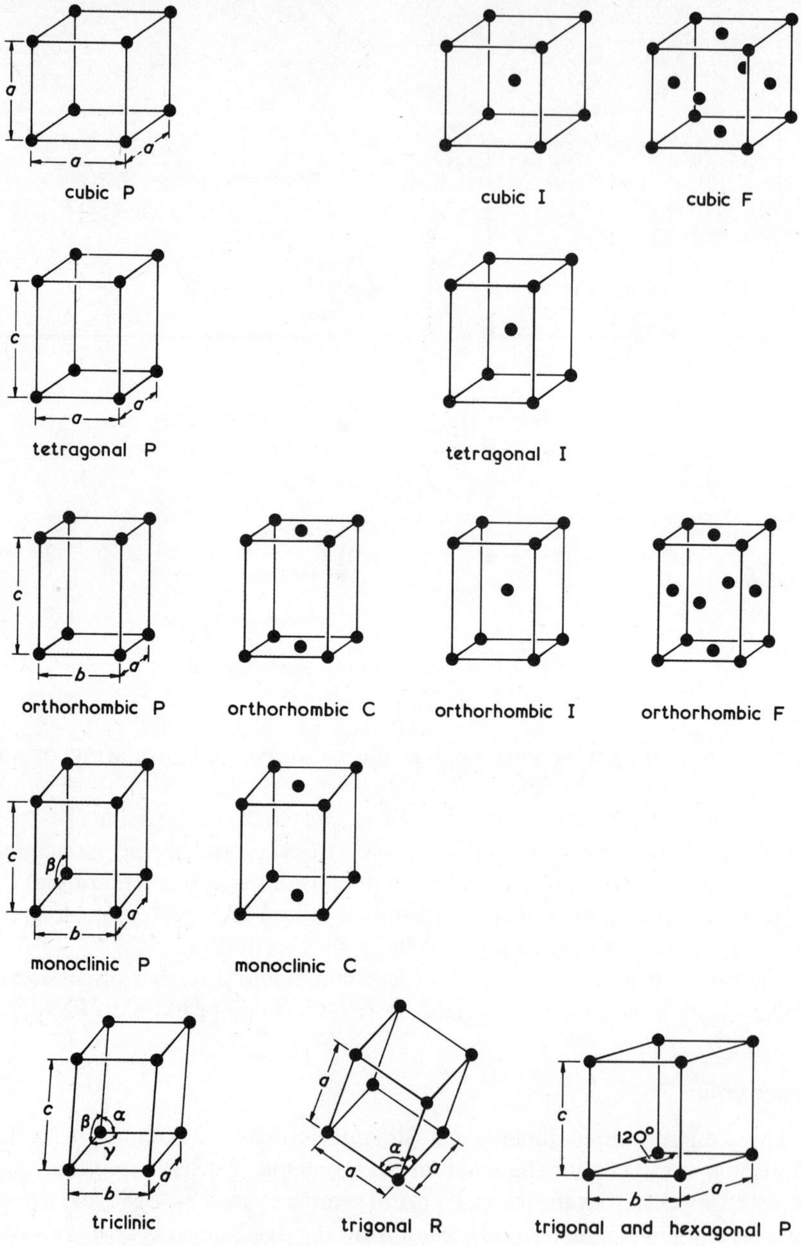

FIG. 2.9 Bravais lattices in three dimensions (angles indicated where $\neq 90°$).

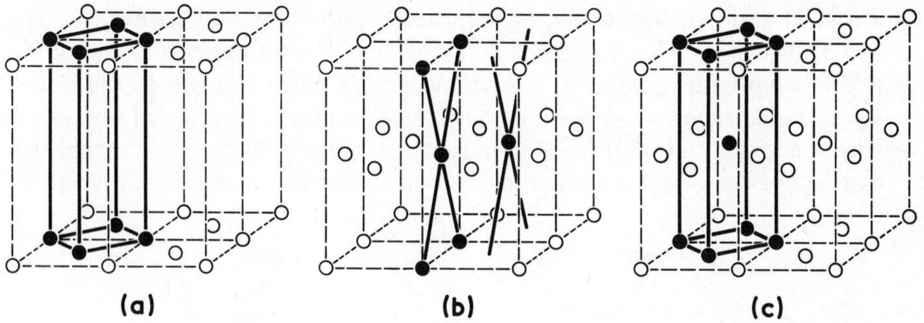

FIG. 2.10 Tetragonal cells.

translational and point groups. There exist 230 space groups, the large number resulting not only from the fact that the point group symmetry of the basis may be lower than that of the Bravais lattice but also that combined translation-rotation and translation-reflexion operations may be allowed. It is worthwhile to describe briefly these operations. *Screw axes* are associated with rotation operations about a 2-, 3-, 4- or 6-fold axis combined with a translation through fractions (which must be multiples of 1/2, 1/3, 1/4, and 1/6 respectively) of the primitive translation. For example the elements 2_1 and 3_1 are indicated in Fig. 2.11. The subscript 1 gives the fractional order (1/2 and

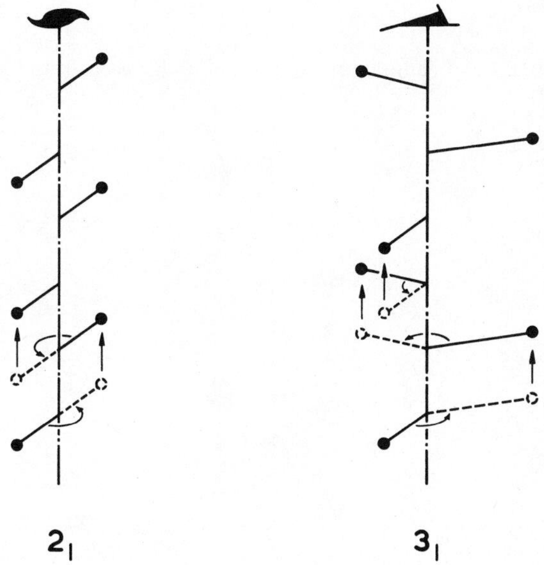

FIG. 2.11 Screw axes 2_1 and 3_1.

1/3 in 2_1 and 3_1 respectively) of the translation operation. The element 3_2 is illustrated in Fig. 2.12 together with its interesting enantiomorphic relationship with the axis 3_1 (i.e. equivalent to spiral rotation in opposite sense). *Glide planes* are associated with the combined operation of reflexion across a plane followed by a translation parallel to the plane through a distance of half of a primitive translation. This element is shown in Fig. 2.13.

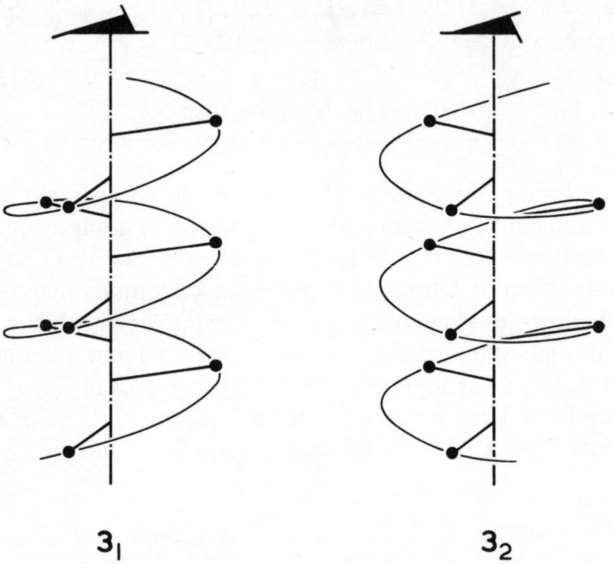

FIG. 2.12 Enantiomorphic relationship between 3_1 and 3_2.

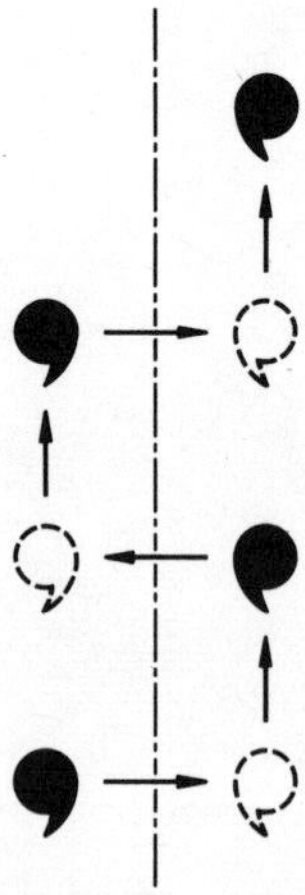

FIG. 2.13 The glide plane.

Internationally agreed symbols for space group operations may be found in the *International Tables for Crystallography* which should be consulted.

Space group notation

Rather than spell out the rules, we will take several examples of space groups, explain the significance of the symbols and represent the space group in two dimensions.

Pbca: the letters *a*, *b* and *c* refer to glide planes, so that we may consider this space group to be derived from the point group *mmm*. The crystal must therefore belong to the orthorhombic system. The *P* understandably implies a primitive lattice. In space group language the *axes* ***a b*** and ***c*** constitute a right handed set, their conventional arrangement illustrated in Fig. 2.14. The order of the symbols *a*, *b* and *c* is very important and implies the following operations

(*i*) *b*: reflect in plane perpendicular to ***a*** and translate in direction ***b***
(*ii*) *c*: reflect in plane perpendicular to ***b*** and translate in direction ***c***
(*iii*) *a*: reflect in plane perpendicular to ***c*** and translate in direction ***a***

We illustrate these operations in Fig. 2.14. Taking a general point indicated by an open circle followed by the sign + to indicate its position above the *ab* plane, we carry out the appropriate operations (*i*), (*ii*) and (*iii*). Reflexions are indicated by the symbol ⊙ and translations in the ***c*** direction, are indicated $\pm 1/2$.

Pnnn: we introduce this space group as a simple extension to *Pbca* to which it is clearly related. The significance of the symbols *n* is as follows

(*i*) *n*: reflect in plane perpendicular to ***a*** and translate 1/2 in direction ***b*** followed by 1/2 in direction ***c***

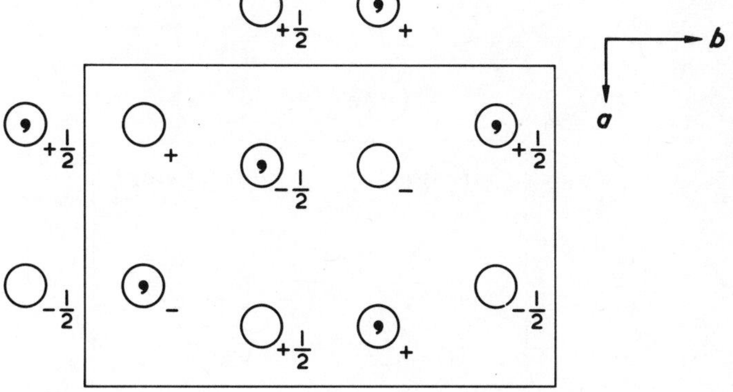

FIG. 2.14 Space group *Pbca*.

(*ii*) *n*: reflect in plane perpendicular to ***b*** and translate 1/2 in direction ***a*** followed by 1/2 in direction ***c***.

(*iii*) *n*: reflect in plane perpendicular to ***c*** and translate 1/2 in direction ***a*** followed by 1/2 in direction ***b***.

The result is illustrated in Fig. 2.15.

$P2_1/c$: derived from the point group $2/m$ of the monoclinic system, the symbols should be self-explanatory. We have to effect the operations 2_1 (conventionally ***b***) on our general point and the ***c***-glide after reflexion in a plane perpendicular to 2_1. The result is seen in Fig. 2.16.

The examples we have chosen are particularly simple, but they illustrate at least the basic technique. The International Tables must be consulted and further experience must be gained in the interpretation of space group symbols.

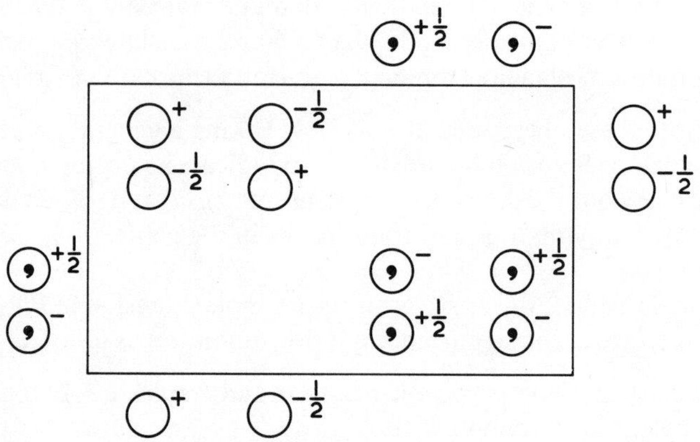

FIG. 2.15 Space group *Pnnn*.

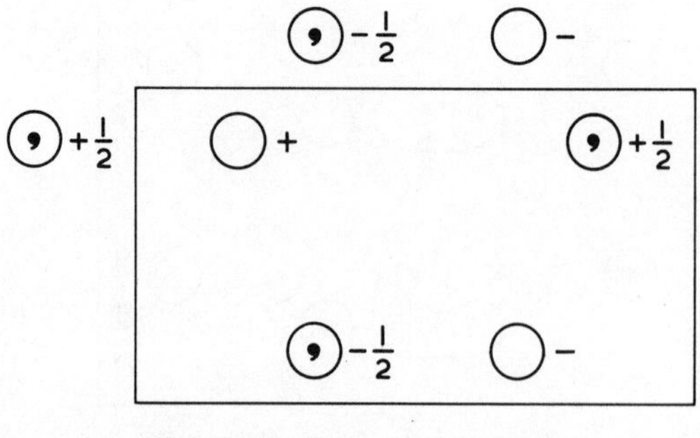

FIG. 2.16 Space group $P2_1/c$.

References

(1) J. F. Nye, *The Physical Properties of Crystals* (Clarendon Press, 1957)
(2) W. Friedrich, P. Knipping and M. von Laue, *Munich Ber.*, 1912, 30; reprinted with slight alterations in *Ann Phys.* **41**, 971 (1913)

General

J. F. Nye (*see* (1))
S. Bhagavantam, *Crystal Symmetry and Physical Properties* (Academic Press, 1966)

3

The Reciprocal Lattice and Crystal Diffraction

Our ultimate aims in this short but very important chapter are to become familiar with the reciprocal lattice and the elementary treatment of crystal diffraction. Before beginning to discuss the former topic, let us ensure we understand Miller indices.

3.1 Miller indices

In solid state physics it is a common requirement to refer to a specific set of *lattice points* which may be considered to make up a *lattice plane*. Indeed it is easy to see that a lattice may be divided into sets of parallel lattice planes whose orientation with respect to the crystal axes may be specified by reference to their Miller indices which may be defined as follows. Let us assume that the lattice plane in question makes intercepts of $m_1 a_1$, $m_2 a_2$ and $m_3 a_3$ on the axes defining the crystal system. The Miller indices are then given by the three numbers $l_1 l_2$ and l_3, which are the smallest integers satisfying the condition

$$l_1 : l_2 : l_3 = \frac{1}{m_1} : \frac{1}{m_2} : \frac{1}{m_3}$$

The Miller indices of a set of planes are always written as $(l_1 l_2 l_3)$ and when one or more intercepts are negative, as $(\bar{l}_1 l_2 l_3)$. When the indices are enclosed in braces i.e. $\{l_1 l_2 l_3\}$ this always refers to a set of symmetry related planes.

3.2 The reciprocal lattice and Brillouin zones

The reciprocal lattice assumes a unique role in the formulation of solid state properties and it is necessary to become as familiar with its concept as with that of the real crystal lattice. The effort required will be quickly rewarded. For example, the Bragg condition for constructive interference takes on an unfamiliar elegance when expressed in language appropriate to reciprocal space. The definition of reciprocal lattice is quite straightforward provided a familiarity with vector products may be assumed.

If $\boldsymbol{a}_1$, $\boldsymbol{a}_2$ and $\boldsymbol{a}_3$ are the three *primitive* translation vectors of the *real* lattice (we need to stress that we refer to *direct* or *real* space), thus we may define the fundamental vectors of the *reciprocal* lattice ($\boldsymbol{b}_1$, $\boldsymbol{b}_2$ and $\boldsymbol{b}_3$) by the equations*

$$\begin{array}{lll} \boldsymbol{a}_1 \cdot \boldsymbol{b}_1 = 2\pi; & \boldsymbol{a}_2 \cdot \boldsymbol{b}_1 = 0; & \boldsymbol{a}_3 \cdot \boldsymbol{b}_1 = 0 \\ \boldsymbol{a}_1 \cdot \boldsymbol{b}_2 = 0 & \boldsymbol{a}_2 \cdot \boldsymbol{b}_2 = 2\pi; & \boldsymbol{a}_3 \cdot \boldsymbol{b}_2 = 0 \\ \boldsymbol{a}_1 \cdot \boldsymbol{b}_3 = 0 & \boldsymbol{a}_2 \cdot \boldsymbol{b}_3 = 0 & \boldsymbol{a}_3 \cdot \boldsymbol{b}_3 = 2\pi \end{array} \tag{3.1}$$

Equations 3.1 may be summarised in an obvious way by use of the Kronecker delta. Thus we may write

$$\boldsymbol{a}_i \cdot \boldsymbol{b}_j = 2\pi\delta_{ij} \qquad (i, j = 1, 2, 3) \tag{3.2}$$

Inspection of 3.2 will show that, for example, since

$$\boldsymbol{a}_1 \cdot \boldsymbol{b}_2 = \boldsymbol{a}_3 \cdot \boldsymbol{b}_2 = 0$$

the reciprocal lattice vector $\boldsymbol{b}_2$ is perpendicular to the plane defined by the direct lattice vectors $\boldsymbol{a}_1$ and $\boldsymbol{a}_3$. It follows from elementary vector algebra that $\boldsymbol{b}_3$ is parallel to the vector $\boldsymbol{a}_3 \times \boldsymbol{a}_1$, or the vectors may be equated by a scalar constant j. Thus

$$\boldsymbol{b}_2 = j(\boldsymbol{a}_3 \times \boldsymbol{a}_1) \tag{3.3}$$

But since we know that $\boldsymbol{a}_2 \cdot \boldsymbol{b}_2 = 2\pi$, it follows that

$$2\pi = j(\boldsymbol{a}_2 \cdot \boldsymbol{a}_3 \times \boldsymbol{a}_1) \tag{3.4}$$

Hence (and similarly for $\boldsymbol{b}_1$ and $\boldsymbol{b}_3$)

$$\boldsymbol{b}_1 = 2\pi \frac{\boldsymbol{a}_2 \times \boldsymbol{a}_3}{\boldsymbol{a}_1 \cdot \boldsymbol{a}_2 \times \boldsymbol{a}_3} \tag{3.5}$$

$$\boldsymbol{b}_2 = 2\pi \frac{\boldsymbol{a}_3 \times \boldsymbol{a}_1}{\boldsymbol{a}_2 \cdot \boldsymbol{a}_3 \times \boldsymbol{a}_1} \tag{3.6}$$

and

$$\boldsymbol{b}_3 = 2\pi \frac{\boldsymbol{a}_1 \times \boldsymbol{a}_2}{\boldsymbol{a}_3 \cdot \boldsymbol{a}_1 \times \boldsymbol{a}_2} \tag{3.7}$$

It should be noted that

$$\boldsymbol{a}_1 \cdot \boldsymbol{a}_2 \times \boldsymbol{a}_3 = \boldsymbol{a}_2 \cdot \boldsymbol{a}_3 \times \boldsymbol{a}_1 = \boldsymbol{a}_3 \cdot \boldsymbol{a}_1 \times \boldsymbol{a}_2 (=?) \tag{3.8}$$

where the significance of (?) may be determined by the reader.

* The reader may frequently find the factor 2π omitted (usually by crystallographers). Solid state physicists normally include it.

It is of pivotal importance to recognise the relationship between the real lattice and its reciprocal analogue. Lattice points in both spaces are evidently defined in an identical way provided one uses the appropriate fundamental vectors.

$$\boldsymbol{R}(n_1 n_2 n_3) = n_1 \boldsymbol{a}_1 + n_2 \boldsymbol{a}_2 + n_3 \boldsymbol{a}_3 \qquad (n_1, n_2 \text{ and } n_3 \text{ are integers}) \tag{3.9}$$

$$\boldsymbol{G}(l_1 l_2 l_3) = l_1 \boldsymbol{b}_1 + l_2 \boldsymbol{b}_2 + l_3 \boldsymbol{b}_3 \qquad (l_1, l_2 \text{ and } l_3 \text{ are integers}) \tag{3.10}$$

Two important relationships may be demonstrated readily. From the previous definition of Miller indices, we note that the points defined by the vectors

$$\frac{\boldsymbol{a}_1}{l_1}, \quad \frac{\boldsymbol{a}_2}{l_2} \quad \text{and} \quad \frac{\boldsymbol{a}_3}{l_3}$$

must be in the plane $(l_1 l_2 l_3)$ so that vectors such as $(\boldsymbol{a}_1/l_1 - \boldsymbol{a}_2/l_2)$ and $(\boldsymbol{a}_1/l_1 - \boldsymbol{a}_3/l_3)$ must also be in this plane. Let us consider the scalar product

$$\begin{aligned} \boldsymbol{G}(l_1 l_2 l_3) \cdot \left(\frac{\boldsymbol{a}_1}{l_1} - \frac{\boldsymbol{a}_2}{l_2}\right) &= (l_1 \boldsymbol{b}_1 + l_2 \boldsymbol{b}_2 + l_3 \boldsymbol{b}_3) \cdot \left(\frac{\boldsymbol{a}_1}{l_1} - \frac{\boldsymbol{a}_2}{l_2}\right) \\ &= 2\pi\left(\frac{l_1}{l_1} - \frac{l_2}{l_2}\right) \\ &= 0 \end{aligned} \tag{3.11}$$

Thus the reciprocal lattice vector $\boldsymbol{G}(l_1 l_2 l_3)$ *is perpendicular to the lattice plane* $(l_1 l_2 l_3)$ *of the direct crystal lattice.*

Now let us assume that the spacing of the planes $(l_1 l_2 l_3)$ is $d(l_1 l_2 l_3)$. We know that the unit vector $\hat{\boldsymbol{G}}$ in the direction $\boldsymbol{G}(l_1 l_2 l_3)$ is normal to the plane $(l_1 l_2 l_3)$. So a general vector $\boldsymbol{r}$ defining the plane is related to d and $\hat{\boldsymbol{G}}$ by the equation

$$\hat{\boldsymbol{G}} \cdot \boldsymbol{r} = d(l_1 l_2 l_3) \tag{3.12}$$

[since the projection of $\boldsymbol{r}$ on $\hat{\boldsymbol{G}}d$ must give the magnitude $d(l_1 l_2 l_3)$]. If we choose, for example, $\boldsymbol{r} = \boldsymbol{a}_1/l_1$ in the $(l_1 l_2 l_3)$ plane, we may write for 3.12.

$$\frac{\boldsymbol{G}}{G} \cdot \left(\frac{\boldsymbol{a}_1}{l_1}\right) = d(l_1 l_2 l_3) \tag{3.13}$$

but since

$$\boldsymbol{G} \cdot \boldsymbol{a}_1 = 2\pi l_1 \tag{3.14}$$

from 3.13 and 3.14

$$d(l_1 l_2 l_3) = \frac{2\pi}{G(l_1 l_2 l_3)} \tag{3.15}$$

(a very important result)

One further important relationship between the real and reciprocal lattice is seen by considering the scalar product $\boldsymbol{G} \cdot \boldsymbol{R}$:

$$\begin{aligned} \boldsymbol{G} \cdot \boldsymbol{R} &= (l_1 \boldsymbol{b}_1 + l_2 \boldsymbol{b}_2 + l_3 \boldsymbol{b}_3) \cdot (n_1 \boldsymbol{a}_1 + n_2 \boldsymbol{a}_2 + n_3 \boldsymbol{a}_3) \\ &= 2\pi(l_1 n_1 + l_2 n_2 + l_3 n_3) \\ &= 2\pi \times \text{an integer} \end{aligned} \tag{3.16}$$

Hence

$$\exp(i\boldsymbol{G} \cdot \boldsymbol{R}) = 1 \tag{3.17}$$

The value of this relationship will become obvious in Chapter 6.

As a simple example, let us consider the reciprocal to the simple *fcc* lattice (see Fig. 3.1). The primitive translation vectors are

$$\boldsymbol{a}_1 = \frac{a}{2}(\hat{\boldsymbol{x}}_1 + \hat{\boldsymbol{x}}_2) \tag{3.18}$$

$$\boldsymbol{a}_2 = \frac{a}{2}(\hat{\boldsymbol{x}}_2 + \hat{\boldsymbol{x}}_3) \tag{3.19}$$

$$\boldsymbol{a}_3 = \frac{a}{2}(\hat{\boldsymbol{x}}_1 + \hat{\boldsymbol{x}}_3) \tag{3.20}$$

Hence the primitive translation vectors of the reciprocal lattice may be evaluated according to previous definitions 3.5–3.7. As an example to make

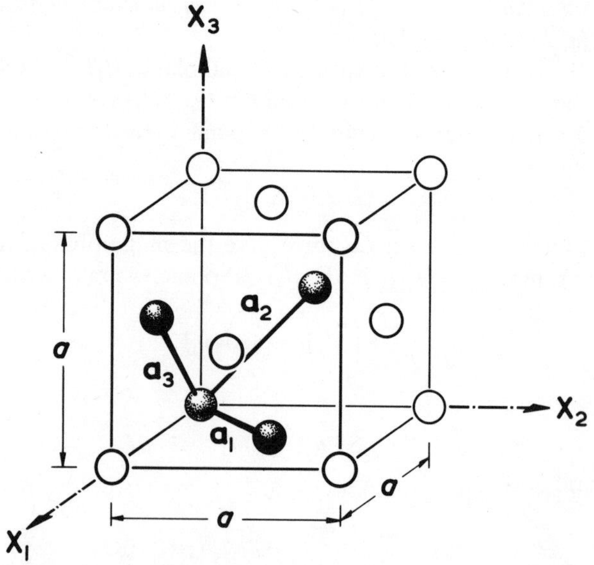

FIG. 3.1 Primitive translation vectors of the *fcc* lattice.

the procedure absolutely clear, let us obtain an expression for $\boldsymbol{b}_1$. From eqns 3.5 and 3.18–3.20

$$\boldsymbol{b}_1 = \frac{4\pi}{a}\frac{(\hat{\boldsymbol{x}}_2 + \hat{\boldsymbol{x}}_3) \times (\hat{\boldsymbol{x}}_1 + \hat{\boldsymbol{x}}_3)}{(\hat{\boldsymbol{x}}_1 + \hat{\boldsymbol{x}}_2)\cdot(\hat{\boldsymbol{x}}_2 + \hat{\boldsymbol{x}}_3) \times (\hat{\boldsymbol{x}}_1 + \hat{\boldsymbol{x}}_3)}$$

$$= \frac{4\pi}{a}\frac{(\hat{\boldsymbol{x}}_1 + \hat{\boldsymbol{x}}_2 - \hat{\boldsymbol{x}}_3)}{(\hat{\boldsymbol{x}}_1 + \hat{\boldsymbol{x}}_2)\cdot(\hat{\boldsymbol{x}}_1 + \hat{\boldsymbol{x}}_2 - \hat{\boldsymbol{x}}_3)}$$

$$= \frac{2\pi}{a}(\hat{\boldsymbol{x}}_1 + \hat{\boldsymbol{x}}_2 - \hat{\boldsymbol{x}}_3) \tag{3.21}$$

Similarly

$$\boldsymbol{b}_2 = \frac{2\pi}{a}(-\hat{\boldsymbol{x}}_1 + \hat{\boldsymbol{x}}_2 + \hat{\boldsymbol{x}}_3) \tag{3.22}$$

and

$$\boldsymbol{b}_3 = \frac{2\pi}{a}(\hat{\boldsymbol{x}}_1 - \hat{\boldsymbol{x}}_2 + \hat{\boldsymbol{x}}_3) \tag{3.23}$$

(It should be confirmed that these are primitive vectors of a *bcc* lattice.) From 3.21–3.23 we may then write the general reciprocal lattice vector

$$\boldsymbol{G} = l_1\boldsymbol{b}_1 + l_2\boldsymbol{b}_2 + l_3\boldsymbol{b}_3$$

$$= \frac{2\pi}{a}(l_1 - l_2 + l_3)\hat{\boldsymbol{x}}_1 + (l_1 + l_2 - l_3)\hat{\boldsymbol{x}}_2 + (-l_1 + l_2 + l_3)\hat{\boldsymbol{x}}_3 \tag{3.24}$$

As is the practice in real space, we may divide the reciprocal lattice into its constituent unit cells and it is appropriate to choose a Wigner–Seitz primitive cell, which may be defined in real or reciprocal space by planes which perpendicularly bisect lines joining adjacent lattice points. We show such a plane in reciprocal space in Fig. 3.2. This plane may be defined in terms of

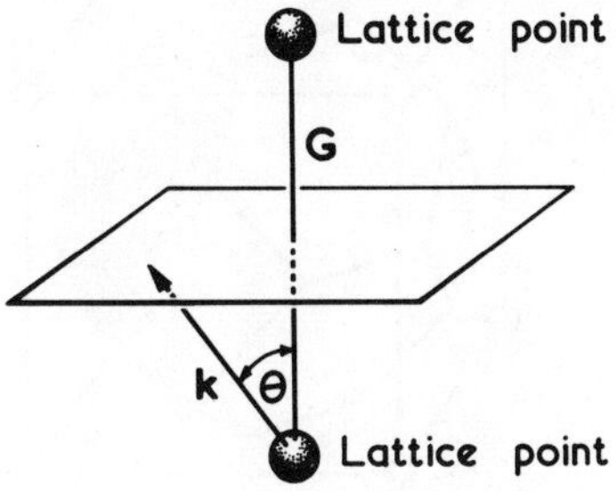

FIG. 3.2 A plane of the Wigner–Seitz cell in reciprocal space.

the reciprocal lattice vector $\boldsymbol{G}$ and some general vector $\boldsymbol{k}$ by means of the equation

$$\boldsymbol{k}\cdot\boldsymbol{G} = kG\cos\theta$$

$$= \frac{G^2}{2} \tag{3.25}$$

or

$$2\boldsymbol{k}\cdot\boldsymbol{G} = G^2 \tag{3.26}$$

It is now a simple matter to construct the Wigner–Seitz cell in reciprocal (or $\boldsymbol{k}$) space. It is known as the first *Brillouin zone.* If we substitute orthogonal components $k_1\hat{\boldsymbol{x}}_1$, $k_2\hat{\boldsymbol{x}}_2$ and $k_3\hat{\boldsymbol{x}}_3$ for the vector $\boldsymbol{k}$ and the expression 3.24 for $\boldsymbol{G}$ in eqn. 3.26 we obtain

$$k_1(l_1 - l_2 + l_3) + k_2(l_1 + l_2 - l_3) + k_3(-l_1 + l_2 + l_3)$$
$$= \frac{\pi}{a}[(l_1 - l_2 + l_3)^2 + (l_1 + l_2 - l_3)^2 + (-l_1 + l_2 + l_3)^2] \tag{3.27}$$

The first Brillouin zone is defined by fourteen bounding planes, obtained by permuting the two sets of l values $(\pm 1 \quad 0 \quad 0)$ and $(\pm 1 \quad \pm 1 \quad 0)$. The equations to the faces are (from 3.27):

$$\pm k_1 \pm k_2 \pm k_3 = \frac{3\pi}{a} \tag{3.28}$$

and

$$\pm k_1 = \frac{2\pi}{a}; \quad \pm k_2 = \frac{2\pi}{a}; \quad \pm k_3 = \frac{2\pi}{a} \tag{3.29}$$

These boundary planes define a truncated octahedron as shown in Fig. 3.3.

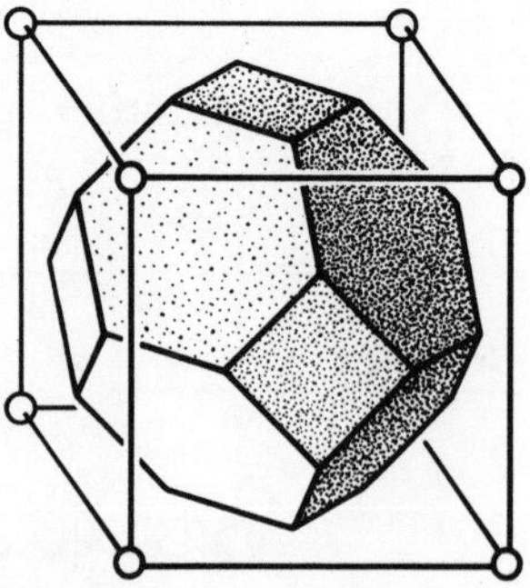

FIG. 3.3 The first Brillouin zone of the *fcc* lattice.

Having been through this calculation in detail, no difficulties should be experienced in obtaining the reciprocal lattice and first Brillouin zone for the *bcc* and hexagonal lattices; this exercise is recommended.

Matrix representation of the reciprocal lattice

Finally before discussing diffraction, we present a slightly different approach to obtaining the primitive translation vectors of the reciprocal lattice. Some readers who decide to read this section, which can be skipped, may find the matrix approach easier than the traditional analysis presented above.

We know that lattice points in direct space may be defined by the vectors

$$\boldsymbol{R} = n_1\boldsymbol{a}_1 + n_2\boldsymbol{a}_2 + n_3\boldsymbol{a}_3$$

Now in order to specify the primitive translation vectors $\boldsymbol{a}_i$ ($i = 1, 2$ or 3), they may be referred to a rectangular coordinate system so that the components of $\boldsymbol{a}_i$ will be a_{ix}, a_{iy} and a_{iz} (here to avoid confusion we are forced to adopt the more traditional labels x, y, and z). These nine components completely specify the lattice and may be conveniently written in matrix form, together with n_1, n_2 and n_3 as a column vector. Each lattice point is then specified by the matrix product

$$\boldsymbol{An} = \begin{pmatrix} a_{1x} & a_{2x} & a_{3x} \\ a_{1y} & a_{2y} & a_{3y} \\ a_{1z} & a_{2z} & a_{3z} \end{pmatrix}\begin{pmatrix} n_1 \\ n_2 \\ n_3 \end{pmatrix} \tag{3.30}$$

It may then be confirmed that matrix $\boldsymbol{A}$ for the simple cubic and *fcc* lattices (referring to Fig. 3.1) are as follows

$$\boldsymbol{A}_{sc} = \begin{pmatrix} a & 0 & 0 \\ 0 & a & 0 \\ 0 & 0 & a \end{pmatrix}; \qquad \boldsymbol{A}_{fcc} = \begin{pmatrix} \frac{a}{2} & \frac{a}{2} & 0 \\ 0 & \frac{a}{2} & \frac{a}{2} \\ \frac{a}{2} & 0 & \frac{a}{2} \end{pmatrix}$$

It is a natural extension to treat reciprocal lattice vectors a similar way such that

$$\boldsymbol{lB} = (l_1 l_2 l_3)\begin{pmatrix} b_{1x} & b_{1y} & b_{1z} \\ b_{2x} & b_{2y} & b_{2z} \\ b_{3x} & b_{3y} & b_{3z} \end{pmatrix} \tag{3.31}$$

represents any point in the reciprocal lattice. The reason for the slight difference of treatment, if not already obvious, will become so when it is recognized that we can write our definition of reciprocal lattice (3.2) in a very convenient way using 3.30 and 3.31.

$$\boldsymbol{BA} = 2\pi \boldsymbol{1}$$

(where $\boldsymbol{1}$ is the unit matrix). For those familiar with the rows and columns manipulation of matrix multiplication (everyone should be after Chapter 2), the $\boldsymbol{B}$ matrix may be written down easily. Thus for *fcc*

$$\boldsymbol{B}_{fcc} = 2\pi \begin{pmatrix} \frac{1}{a} & -\frac{1}{a} & \frac{1}{a} \\ \frac{1}{a} & \frac{1}{a} & -\frac{1}{a} \\ -\frac{1}{a} & \frac{1}{a} & \frac{1}{a} \end{pmatrix} \tag{3.32}$$

which is really a description identical with that of eqns 3.21–3.23.

3.3 Crystal diffraction

It will be assumed that the reader is familiar with the elementary physics of diffraction and the simple trigonometric statement of the Bragg law so that the following treatment does not specifically deal in detail with these topics. Further, since we are not too concerned in this book with methods of structure determination (chemists should know about this anyway) the latter are not mentioned. Rather, we wish to gain some general (if superficial) feeling for the importance of diffraction effects in crystals, so it must be admitted that the following discussion will not serve other than as an introduction to standard texts on diffraction.

We may represent the amplitude (ϕ) of a plane wave of frequency ω in free space at some vectorial position $\boldsymbol{r}$ by the equation

$$\phi(\boldsymbol{r}) = \phi_0 \exp[i(\boldsymbol{k}_0 \cdot \boldsymbol{r} - \omega t)] \tag{3.33}$$

where the vector $\boldsymbol{r}$ is defined by choice of an origin at $\boldsymbol{r} = 0$ and where $\boldsymbol{k}_0$ is the familiar wave vector whose magnitude k is related to the wavelength of the radiation by

$$k = \frac{2\pi}{\lambda} \tag{3.34}$$

We may consider the diffraction of the radiation by an array of point scatterers. For convenience we eliminate the time dependence from 3.33 and consider first a scatterer at the origin. If the amplitude of the scattered radiation is observed at $\boldsymbol{A}$ (A is large), we may write

$$\phi(\boldsymbol{A}) = \frac{\phi_0 \alpha}{A} \exp(ikA) \tag{3.35}$$

where k is unchanged if we assume elastic scattering and α is a factor describing the strength of the scattering. The reciprocal dependence of $\phi(\boldsymbol{A})$ on A is consistent with the intensity (I) of the scattered radiation being proportional to A^{-2} [since $I \propto \phi^*(\boldsymbol{A})\phi(\boldsymbol{A})$].

If the scatterer is not at the origin but at position $\boldsymbol{r}$, then an extra phase factor must be introduced to allow for the difference in path lengths. We note that since A is large, paths from $\boldsymbol{r}$ and from the origin may be assumed parallel. If the wave vector of the scattered radiation is $\boldsymbol{k}_1$, the usual simple geometric construction will show that the phase difference is $\boldsymbol{r} \cdot \boldsymbol{k}_1 - \boldsymbol{r} \cdot \boldsymbol{k}_0$. It is convenient to define a scattering vector $\boldsymbol{K}$ such that

$$\boldsymbol{K} = \boldsymbol{k}_1 - \boldsymbol{k}_0 \tag{3.36}$$

Hence we may write for the amplitude at $\boldsymbol{A}$

$$\phi(\boldsymbol{A}) = \frac{\phi_0 \alpha}{A} \exp(ikA - i\boldsymbol{K} \cdot \boldsymbol{r}) \tag{3.37}$$

Now if we consider the amplitude of scattered radiation at $\boldsymbol{A}$ due to N point scatterers in the crystal, 3.37 becomes

$$\phi_{\text{total}}(\boldsymbol{A}) = \frac{\phi_0 \alpha}{A} \sum_{j=1}^{N} \exp(ikA - i\boldsymbol{K} \cdot \boldsymbol{r}_j) \tag{3.38}$$

where $\boldsymbol{r}_j$ defines the position of each of the N point scatterers.

Therefore

$$\phi_{\text{total}}(\boldsymbol{A}) = \frac{\phi_0 \alpha}{A} \exp(ikA) \sum_{j=1}^{N} \exp(-i\boldsymbol{K} \cdot \boldsymbol{r}_j) \tag{3.39}$$

If instead of considering an assembly of point scatterers, we considered some continuous scattering density $\rho(\boldsymbol{r})$, the following expression, analogous to 3.39 would be obtained

$$\phi_{\text{total}}(\boldsymbol{A}) = \frac{\phi_0 \beta}{A} \exp(ikA) \int \rho(\boldsymbol{r}) \exp(-i\boldsymbol{K} \cdot \boldsymbol{r})\, \mathrm{d}\boldsymbol{r} \tag{3.40}$$

For those familiar with the mathematics, the integral in 3.40 will be recognised immediately as a Fourier integral. Further, 3.39 may also be written as a Fourier transform if we define a scattering density in terms of point delta functions such that

$$\rho(\boldsymbol{r}) = \sum \delta(\boldsymbol{r} - \boldsymbol{r}_j) \tag{3.41}$$

Let us suppose that we may use the $\rho(\boldsymbol{r})$ of eqn. 3.40 to represent the variation in electron density over a single atom, which we assume to be spherically symmetric and of radius r. It is an interesting exercise to show that eqn. 3.40 reduces to the form

$$\phi_{\text{atom}} \propto \int_0^\infty 4\pi r^2 \rho(r) \frac{\sin Kr}{Kr}\, \mathrm{d}r \tag{3.42}$$

The quantity on the right-hand side of eqn. 3.42, usually written $f(K)$ is commonly referred to as the *atomic scattering factor* or *form factor*. This is very useful since it means that instead of atoms of finite size in our crystal, we can replace them by point scatterers of appropriate $f(K)$.

Scattering from a lattice of point atoms

Let us imagine that our crystal contains an identical point scatterer at every lattice point; thus

$$\boldsymbol{R} = \boldsymbol{r} = n_1\boldsymbol{a}_1 + n_2\boldsymbol{a}_2 + n_3\boldsymbol{a}_3 \tag{3.43}$$

We take the array of scatterers to be a finite parallelepiped of dimensions $N_1\boldsymbol{a}_1, N_2\boldsymbol{a}_2, N_3\boldsymbol{a}_3$. An expression for the amplitude of the scattered radiation is

$$\phi_{\text{cryst}} \propto \sum_{\boldsymbol{R}} \exp(-i\boldsymbol{K}\cdot\boldsymbol{R}) \tag{3.44}$$

$$= \sum_{n_1=0}^{N_1-1} \sum_{n_2=0}^{N_2-1} \sum_{n_3=0}^{N_3-1} \exp[-i\boldsymbol{K}\cdot(n_1\boldsymbol{a}_1 + n_2\boldsymbol{a}_2 + n_3\boldsymbol{a}_3)] \tag{3.45}$$

$$= \sum_{n_1=0}^{N_1-1} \exp[-in_1(\boldsymbol{K}\cdot\boldsymbol{a}_1)] \sum_{n_2=0}^{N_2-1} \exp[-in_2(\boldsymbol{K}\cdot\boldsymbol{a}_2)] \sum_{n_3=0}^{N_3-1} \exp[-in_3(\boldsymbol{K}\cdot\boldsymbol{a}_3)] \tag{3.46}$$

Making use of summation equations of the type,

$$\sum_{n=0}^{N-1} x^n = \sum_{n=0}^{\infty} x^n - \sum_{n=N}^{\infty} x^n = \frac{1}{1-x} - \frac{x^n}{1-x} \tag{3.47}$$

it follows that, for example,

$$\sum_{n_1=0}^{N_1-1} \exp[-in_1(\boldsymbol{K}\cdot\boldsymbol{a}_1)] = \frac{1 - \exp[-iN_1(\boldsymbol{K}\cdot\boldsymbol{a}_1)]}{1 - \exp[-i(\boldsymbol{K}\cdot\boldsymbol{a}_1)]} \tag{3.48}$$

$$= \frac{\exp[-\frac{1}{2}iN_1(\boldsymbol{K}\cdot\boldsymbol{a}_1)]}{\exp[-\frac{1}{2}i(\boldsymbol{K}\cdot\boldsymbol{a}_1)]}\left\{\frac{\exp[\frac{1}{2}iN_1(\boldsymbol{K}\cdot\boldsymbol{a}_1)] - \exp[-\frac{1}{2}iN_1(\boldsymbol{K}\cdot\boldsymbol{a}_1)]}{\exp[\frac{1}{2}i(\boldsymbol{K}\cdot\boldsymbol{a}_1)] - \exp[-\frac{1}{2}i(\boldsymbol{K}\cdot\boldsymbol{a}_1)]}\right\} \tag{3.49}$$

Since the expression for the intensity (I) of the scattered radiation is equal to $\phi^*_{\text{cryst}}\phi_{\text{cryst}}$, it must contain the product of three terms each of which is a product of the form of 3.49 with its complex conjugate. Somewhat tedious algebra leads finally to

$$I \propto \left[\frac{\sin^2\frac{1}{2}N_1(\boldsymbol{K}\cdot\boldsymbol{a}_1)}{\sin^2\frac{1}{2}(\boldsymbol{K}\cdot\boldsymbol{a}_1)}\right]\left[\frac{\sin^2\frac{1}{2}N_2(\boldsymbol{K}\cdot\boldsymbol{a}_2)}{\sin^2\frac{1}{2}(\boldsymbol{K}\cdot\boldsymbol{a}_2)}\right]\left[\frac{\sin^2\frac{1}{2}N_3(\boldsymbol{K}\cdot\boldsymbol{a}_3)}{\sin^2\frac{1}{2}(\boldsymbol{K}\cdot\boldsymbol{a}_3)}\right] \tag{3.50}$$

Maxima occur in the expression on the right-hand side of 3.50 when the equations

$$\boldsymbol{K}\cdot\boldsymbol{a}_1 = 2\pi l_1 \tag{3.51}$$

$$\boldsymbol{K}\cdot\boldsymbol{a}_2 = 2\pi l_2 \tag{3.52}$$

$$\boldsymbol{K}\cdot\boldsymbol{a}_3 = 2\pi l_3 \tag{3.53}$$

are satisfied simultaneously for integral values of l_1, l_2 and l_3. The eqns 3.51–3.53 are known as the *Laue equations* and may be shown to be equivalent to the Bragg law. The Laue equations find simple expression if we utilise the concept of reciprocal lattice space. First they may be usefully combined after multiplying each by a suitable integer.

$$n_1\boldsymbol{K}\cdot\boldsymbol{a}_1 = 2\pi l_1 n_1 \tag{3.54}$$

$$n_2\boldsymbol{K}\cdot\boldsymbol{a}_2 = 2\pi l_2 n_2 \tag{3.55}$$

$$n_3\boldsymbol{K}\cdot\boldsymbol{a}_3 = 2\pi l_3 n_3 \tag{3.56}$$

Hence

$$\boldsymbol{K}\cdot(n_1\boldsymbol{a}_1 + n_2\boldsymbol{a}_2 + n_3\boldsymbol{a}_3) = 2\pi \quad (x \text{ integer})$$

or

$$\boldsymbol{K}\cdot\boldsymbol{R} = 2\pi \quad (x \text{ integer}) \tag{3.57}$$

Thus by comparison with 3.16, we see that the Laue conditions are satisfied if

$$\boldsymbol{K} = \boldsymbol{G} \tag{3.58}$$

This beautiful result is extremely important; in the present context for example, it means that the familiar diffraction pattern is no more than a map of the reciprocal lattice. Further, since we know already that the separation of lattice planes is related to the reciprocal lattice vector by 3.15 and a simple geometric construction involving $\boldsymbol{k}_0$, $\boldsymbol{k}_1$ and $\boldsymbol{K}$ will relate K to the scattering angle (see following paragraph), it is not too difficult to derive the Bragg law. This exercise is left for the reader to complete.

The Ewald sphere

Here we refer to a geometric construction first used by P. P. Ewald which results in a statement of the diffraction law in terms of the reciprocal lattice. The points of Fig. 3.4 are reciprocal lattice points. The incident radiation

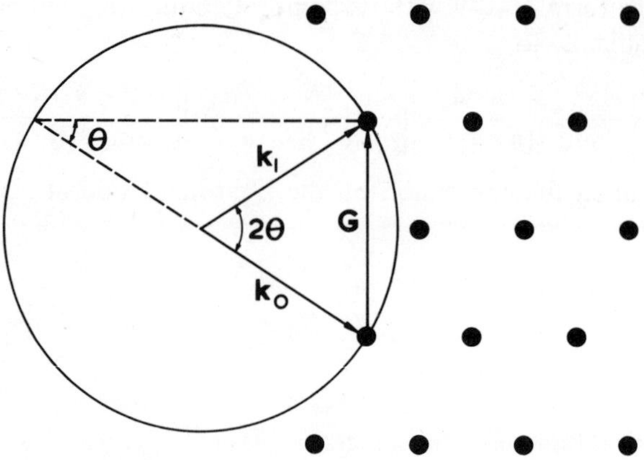

FIG. 3.4 The Ewald sphere construction.

wave vector ($\boldsymbol{k}_0$) is drawn to intersect with one of the latter. A sphere of radius $k(=2\pi/\lambda)$ is drawn about the origin of $\boldsymbol{k}_0$ and constructive interference will be observed if the sphere intersects with any other reciprocal lattice point. Thus in Fig. 3.4, the diffraction condition $\boldsymbol{K} = \boldsymbol{G}$ is satisfied. We note from our definition of $\boldsymbol{K}$ (3.36) or from Fig. 3.4 that

$$\boldsymbol{k}_0 + \boldsymbol{G} = \boldsymbol{k}_1 \tag{3.59}$$

On squaring each side of eqn. 3.59 we obtain

$$k_0^2 + 2\boldsymbol{k}_0 \cdot \boldsymbol{G} + G^2 = k_1^2 \tag{3.60}$$

But since we assume elastic scattering, $k_0 = k_1$, so that eqn. 3.60 is equivalent to

$$2\boldsymbol{k}_0 \cdot \boldsymbol{G} + G^2 = 0 \tag{3.61}$$

a form of the diffraction law to which we refer again. (To whet the appetite of the reader, the intriguing identity of eqns 3.26 and 3.61 could be pondered.)

The structure factor

Our expression of the diffraction condition by 3.58 indicates all possible reflections from a given crystal lattice may be determined. We have made no mention of the nature of the *contents* of the unit cells (or the atomic environment of the lattice points). These, in fact, determine the intensity of individual reflexions. Let us assume there are S atoms per unit cell each specified by the vectors $\boldsymbol{r}_j$ (for the jth atom of atomic scattering factor f_j) whose origin is at a lattice point, i.e.

$$\boldsymbol{r}_j = m_j\boldsymbol{a}_1 + n_j\boldsymbol{a}_2 + p_j\boldsymbol{a}_3 \tag{3.62}$$

Thus eqn. 3.44 may be expanded to give

$$\phi_{\text{cryst}} = \sum_{\boldsymbol{R}} \sum_{j=1}^{S} f_j \exp[-i\boldsymbol{K}\cdot(\boldsymbol{R} + \boldsymbol{r}_j)] \tag{3.63}$$

$$= \sum_{\boldsymbol{R}} \exp(-i\boldsymbol{K}\cdot\boldsymbol{R}) \sum_{d=1}^{s} f_j \exp(-i\boldsymbol{K}\cdot r_j) \tag{3.64}$$

We refer to the right-hand component of the summation product as the *structure factor* $F(\boldsymbol{K})$ which in integral form can be thought of as the Fourier transform of the atomic environment of a lattice point.

$$F(\boldsymbol{K}) = \sum_{j=1}^{S} f_j \exp(-i\boldsymbol{K}\cdot\boldsymbol{r}_j) \tag{3.65}$$

If we now use the diffraction condition 3.58 we see that

$$\boldsymbol{K}\cdot\boldsymbol{r}_j = (m_j\boldsymbol{a}_1 + n_j\boldsymbol{a}_2 + p_j\boldsymbol{a}_3)\cdot(l_1\boldsymbol{b}_1 + l_2\boldsymbol{b}_2 + l_3\boldsymbol{b}_3) \tag{3.66}$$

$$= 2\pi(m_jl_1 + n_jl_2 + p_jl_3) \tag{3.67}$$

So that 3.65 becomes

$$F(l_1l_1l_3) = \sum_{j=1}^{S} f_j \exp[-2\pi i(m_jl_1 + n_jl_2 + p_jl_3)] \tag{3.68}$$

Systematic absences

Equation 3.68 indicates that the structure factor is determined by the position of atoms in the unit cell as indicated by the space group to which the crystal belongs. If this space group is non-primitive or contains glide planes or screw axes, then the phase relationships between scattered radiation from atoms related by these elements is such that the structure factor may be zero for certain reflexions. We illustrate this by reference to the structure factor for the *bcc* lattice.

Referred to an arbitrary origin, the basis of this lattice contains identical atoms at the fractional coordinates

$$[m_j, n_j, p_j] \quad \text{and} \quad [(m_j + \tfrac{1}{2}), (n_j + \tfrac{1}{2}), (p_j + \tfrac{1}{2})].$$

Thus

$$F(l_1 l_2 l_3) = \sum_{j=1}^{S/2} f_j\{\exp[-2\pi i(l_1 m_j + l_2 n_j + l_3 p_j)]$$

$$+ \exp[-2\pi i l_1(m_j + \tfrac{1}{2}) - 2\pi i l_2(n_j + \tfrac{1}{2}) - 2\pi i l_3(p_j + \tfrac{1}{2})]\} \quad (3.69)$$

$$= \sum_{j=1}^{S/2} f_j\{\exp[-2\pi i(l_1 m_j + l_2 n_j + l_3 p_j)]\}$$

$$\times \{1 + \exp[-\pi i(l_1 + l_2 + l_3)]\} \quad (3.70)$$

We now have to consider the conditions under which the structure factor is zero or non-zero. The former condition can only arise when the second component of the summed product is zero. That is when

$$1 + \cos \pi(l_1 + l_2 + l_3) - i \sin \pi(l_1 + l_2 + l_3) = 0 \quad (3.71)$$

This condition is satisfied when $l_1 + l_2 + l_3$ is odd. Correspondingly, when $l_1 + l_2 + l_3$ is even, the left-hand side of eqn. 3.71 equals 2. Thus we would expect spots in a diffraction pattern arising from the planes whose summed Miller indices are odd to be *systematically absent* since under these circumstances the structure factor is zero.

It would be a desirable exercise to carry out an analogous analysis of the *fcc* structure and those readers who are sufficiently interested might enquire into systematic absences from the diffraction patterns of crystals whose space groups contain glide planes or screw axes

References

General

P. J. Brown and J. B. Forsyth, *The Crystal Structure of Solids*, (Edward Arnold, 1974)

D. C. Champeney, *Fourier Transforms and their Physical Applications* (*Techniques of Physics 1*), (Academic Press, 1973)

4

Lattice Specific Heats and Lattice Vibrations

To embark on reading a chapter whose title contains the words specific heats may not seem particularly appealing! Yet this topic has considerable significance both in the history of the quantum theory and the history of solid state physics. Indeed certain aspects of the theory of lattice vibrations and specific heats are still the subject of vigorous research activity, so it is unjustifiable to feel that this subject should be left untouched after elementary school physics.

4.1 The Einstein theory of the specific heat of solids

Let us consider a simple electrically insulating solid containing N_0 identical atoms. Classically we might suppose that with $3N_0$ degrees of freedom and each atom assumed to vibrate independently, the mean total internal energy (U) would be $3N_0 k_B T$, where k_B is Boltzmann's constant and T is the absolute temperature. Hence the specific heat C_v is given by

$$C_v = \left(\frac{\partial U}{\partial T}\right)_{NV} \tag{4.1}$$

$$= 3N_0 k_B \tag{4.2}$$

$$= 3R \text{ (for 1 mole)} \tag{4.3}$$

and should thus be a constant independent of temperature. This simple calculation was first made by Boltzmann and, of course, 'confirms' the law of Dulong and Petit. Indeed at moderate temperatures (say 300–400 K), this law is known to be reasonably well obeyed, the greatest deviations being apparent for light elements. The dramatic limitation of the approach involving the classical equipartition of energy is seen immediately when values of specific heat at low temperatures are investigated: values approach zero at 0 K. This was the situation at the beginning of the century, a time when it seemed that except for the clearing up of one or two loose ends, classical physics had solved *all* problems. One outstanding 'loose end' was, of course, the inability of classical statistical mechanics to account for the energy distribution of emitted thermal radiation. On 14 December 1900 (*annus mirabilis*), Max Planck announced at a meeting of the German Physical Society that the apparent paradox could be resolved if it were assumed that the radiation energy was *discontinuous*, the individual energy elements being labelled *quanta*. Thus occurred the birth of the period which Gamow has called 'the thirty years which shook physics'. Einstein's application[1] of the new theory of the quantum to the specific heat of solids must be regarded as one of the first successes of the theory. He supposed that the solid could be regarded (as was done classically) as an array of atomic oscillators, all vibrating with the same frequency, but that the allowed energy states of these oscillators are integral multiples of $h\nu$ (h = Planck constant; ν = frequency). Einstein found that the application of Boltzmann statistics to this system led to a physical model whose behaviour was consistent with experiment. Let us look in a little more detail at this model; but for variety, let us abandon the familiar treatment given in elementary books on statistical mechanics.

It may be shown that a crystal containing N atoms which interact according to Hook's law is mechanically equivalent to a set of $3N$ independent oscillators. Thus the total internal energy of the crystal is equal to the sum of the mean energies of the individual oscillators. We may define the mean energy ($\bar{\varepsilon}$) of an atomic oscillator by means of the equation

$$\bar{\varepsilon} = \langle n \rangle \hbar\omega \tag{4.4}$$

(here we prefer $\hbar = h/2\pi$ and $\omega = 2\pi\nu$), where $\langle n \rangle$ is the average value of n when the oscillators are in thermal equilibrium. The probability P_i of the non-degenerate state of energy ε_i being occupied is given by the Boltzmann equation

$$P_i = \frac{\exp(-\varepsilon_i/k_B T)}{q} \tag{4.5}$$

where q = the partition function

$$= \sum_{j=0}^{\infty} \exp\left(-\frac{\varepsilon_j}{k_B T}\right) \tag{4.6}$$

The mean energy

$$\langle n \rangle \hbar\omega = \sum P_i \varepsilon_i \tag{4.7}$$

$$= \frac{\sum_{n=0}^{\infty} n\hbar\omega \exp(-n\hbar\omega/k_B T)}{\sum_{n=0}^{\infty} \exp(-n\hbar\omega/k_B T)} \tag{4.8}$$

Thus

$$\langle n \rangle = \frac{\sum_{n=0}^{\infty} n[\exp(-\hbar\omega/k_B T)]^n}{\sum_{n=0}^{\infty} [\exp(-\hbar\omega/k_B T)]^n} \tag{4.9}$$

Equation 4.9 may be simplified readily since the denominator is an infinite power series with a simple summation and the numerator may be summed by noting that it is of the form

$$\sum_{n=0}^{\infty} nx^n = x\frac{\mathrm{d}}{\mathrm{d}x}\sum_{n=0}^{\infty} x^n = \frac{x}{(1-x)^2}$$

Hence

$$\langle n \rangle = \frac{1}{\exp(\hbar\omega/k_B T) - 1} \tag{4.10}$$

(One reason for following this method is that $\langle n \rangle$ is essentially the Bose–Einstein distribution function which will be referred to later when quantum statistics are *very briefly* reviewed.)

Using 4.10 we may write in general an expression for the total internal energy U

$$U = \sum_{i=1}^{3N} \frac{\hbar\omega_i}{\exp(\hbar\omega_i/k_B T) - 1} \tag{4.11}$$

when we assume that the $3N$ oscillators have differing frequencies ω_i ($i = 1, 2, 3, \ldots 3N$). The summation of 4.11 could be replaced by an integral if we could assume a quasi-continuous distribution of frequencies $\mathcal{N}(\omega)$. This is a treatment to which we will become very familiar in Chapter 5 and is very convenient when a large number of energy levels are very closely distributed. Therefore instead of 4.11, we write

$$U = \int_0^{\omega_m} \frac{\hbar\omega}{\exp(\hbar\omega/k_B T) - 1}\mathcal{N}(\omega)\,\mathrm{d}\omega \tag{4.12}$$

where $\mathcal{N}(\omega)\,\mathrm{d}\omega$ is the number of oscillators whose frequencies are in the range ω and $\omega + \mathrm{d}\omega$; ω_m is the maximum frequency of any oscillator. Clearly therefore we have the further equation

$$\int_0^{\omega_m} \mathcal{N}(\omega)\,\mathrm{d}\omega = 3N \tag{4.13}$$

Hence from 4.1, we may derive expressions for C_v using 4.11 or 4.12

$$C_v = \sum_{i=1}^{3N} k_B \left(\frac{\hbar\omega_i}{k_B T}\right)^2 \frac{\exp(\hbar\omega_i/k_B T)}{[\exp(\hbar\omega_i/k_B T) - 1]^2} \tag{4.14}$$

and

$$C_v = \int_0^{\omega_m} k_B \left(\frac{\hbar\omega}{k_B T}\right)^2 \frac{\exp(\hbar\omega/k_B T)}{[\exp(\hbar\omega/k_B T) - 1]^2} \mathcal{N}(\omega)\, d\omega \tag{4.15}$$

It is not immediately apparent on inspection that eqn. 4.14 has the correct general form, but in fact, since the summand is a monotonically increasing factor, the total sum must behave in the same way. It will also be noted that at large values of T ($\hbar\omega_i \ll k_B T$), the summand tends to k_B. (This can be seen on expansion of the expression; thus the classical limit leads to the Dulong and Petit law.)

Thus we have developed what so far appears to be potentially a useful theory but its significance can only be assessed if we are able to choose the most appropriate form of the frequency distribution function. Einstein adopted the simplest approach by postulating that all of the atoms oscillate with an identical frequency ω. If we let $\theta_E = \hbar\omega/k_B$, eqn. 4.14 takes the form

$$C_v = 3Nk_B \left(\frac{\theta_E}{T}\right)^2 \frac{\exp(\theta_E/T)}{[\exp(\theta_E/T) - 1]^2} \tag{4.16}$$

Experimentally, of course, specific heats approach the zero of absolute temperature according to the well-known T^3 law, but, although the Einstein specific heat equation does not contain the T^3 term, nevertheless the agreement with experiment is very encouraging if a 'best-fit' choice of θ_E is made (see Fig. 4.1).

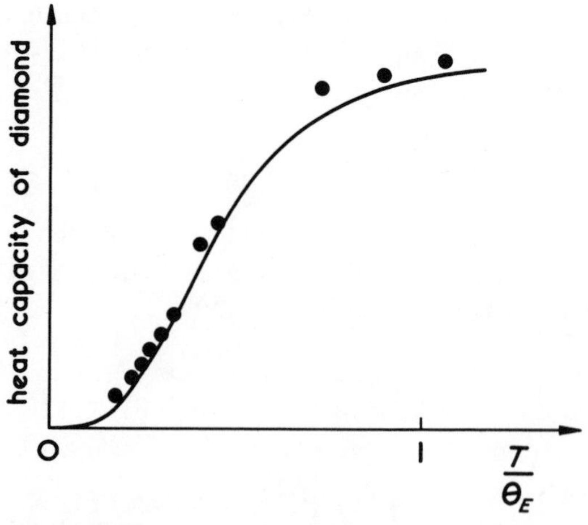

FIG. 4.1 Experimental test of the Einstein specific heat equation.

4.2 The Debye theory of the specific heat of solids

As an alternative to the Einstein model, Debye suggested[2] that a more appropriate distribution function $\mathcal{N}(\omega)$ might be found by treating a solid as an isotropic elastic continuum whose elastic constants are independent of frequency. This appears conceptually satisfactory provided that the wavelengths of the elastic modes are large compared with the atomic separations within the medium. In this sense we might say that the elastic wave 'sees' the material as a continuum without being 'aware' of its discrete atomic constitution (it is difficult to avoid these anthropomorphisms). Debye's assumption was that the *whole range* of atomic vibrational frequencies could be described by the distribution function obtained from the continuum model. In order to get some feel for this problem, let us rapidly develop the model for one- and three-dimensional solids. We will assume without justification the equations describing the elastic waves and certain of their polarisation properties.

One-dimensional solid

We consider a one-dimensional crystal of length L and that only longitudinal (x) displacement occurs. At a time t, we denote the displacements by $u(x, t)$. The partial differential equation describing the displacement is

$$\frac{\partial^2 u}{\partial x^2} = \frac{1}{v_1{}^2}\frac{\partial^2 u}{\partial t^2} \tag{4.17}$$

where v_1 is the velocity of the wave (subscript 1 refers to one-dimension). Assuming the boundary conditions that the solid is fixed at each end (i.e. $u = 0$ at $x = 0$ and $x = L$), the solutions to 4.17 are equations of standing waves.

$$u(x, t) = A \sin\left(\frac{n\pi x}{L}\right)\cos \omega t \tag{4.18}$$

where $n = 1, 2, 3$ etc. and A is a constant

The frequency of the vibrations is given by

$$\omega = \frac{\pi v_1 n}{L} \tag{4.19}$$

By differentiation of eqn. 4.19 we immediately obtain

$$\mathcal{N}(\omega) = \frac{L}{\pi v_1} \tag{4.20}$$

which is a constant independent of frequency.

We have presented this very simple calculation in order to establish the basic approach. We may now progress to the three-dimensional case, which is a little more troublesome.

Three-dimensional isotropic solid

Let us consider for simplicity a solid cube with edges of length L; a point $(x_1 x_2 x_3)$ is displaced to $(x_1 + u_1, x_2 + u_2, x_3 + u_3)$ during the propagation of the elastic wave. If the most general partial differential equation with coefficients determined by the elastic constants is used, it is found that there are two distinguishable velocities of propagation; v_{3l} for the longitudinal wave, a non-degenerate mode in the direction of propagation, and v_{3t} for the transverse waves, doubly degenerate modes perpendicular to the direction of propagation. These velocities are related by elastic constants in the following way

$$v_{3l} = \left[\frac{3(1 - \sigma)}{\beta\rho(1 + \sigma)}\right]^{1/2} \tag{4.21}$$

and

$$v_{3t} = \left[\frac{3(1 - 2\sigma)}{2\beta\rho(1 + \sigma)}\right]^{1/2} \tag{4.22}$$

where β is the bulk compressibility, σ is Poisson's ratio and ρ is the density. We may employ, therefore, two differential equations involving respectively v_{3l} and v_{3t}, each analogous to 4.17

$$\frac{\partial^2 u_l}{\partial {x_1}^2} + \frac{\partial^2 u_l}{\partial {x_2}^2} + \frac{\partial^2 u_l}{\partial {x_3}^2} = \frac{1}{v_{3l}^2}\frac{\partial^2 u_l}{\partial_t^{\,2}} \tag{4.23}$$

and

$$\frac{\partial^2 u_t}{\partial {x_1}^2} + \frac{\partial^2 u_t}{\partial {x_2}^2} + \frac{\partial^2 u_t}{\partial {x_3}^2} = \frac{1}{v_{3t}^2}\frac{\partial^2 u_t}{\partial t^2} \tag{4.24}$$

Choosing again the simplest boundary conditions that the surface planes are held rigidly, we obtain the following standing wave solutions

$$u_l = A_l \sin\left(\frac{\pi n_1 x_1}{L}\right)\sin\left(\frac{\pi n_2 x_2}{L}\right)\sin\left(\frac{\pi n_3 x_3}{L}\right)\cos \omega_l t \tag{4.25}$$

and

$$u_t = A_t \sin\left(\frac{\pi n_1 x_1}{L}\right)\sin\left(\frac{\pi n_2 x_2}{L}\right)\sin\left(\frac{\pi n_3 x_3}{L}\right)\cos \omega_t t \tag{4.26}$$

where n_1, n_2 and n_3 are arbitrary integers and A_l and A_t are constants. The integers n_1, n_2 and n_3 are related to the frequencies by the equations

$$\omega_l^2 = \frac{\pi^2 v_l^2}{L^2}(n_1^2 + n_2^2 + n_3^2) \tag{4.27}$$

and

$$\omega_t^2 = \frac{\pi^2 v_t^2}{L^2}(n_1^2 + n_2^2 + n_3^2) \tag{4.28}$$

Thus for any chosen set of integers n_1, n_2 and n_3 there are three independent modes of vibration (one longitudinal and two transverse).

The wavelengths of the modes 4.25 and 4.26 measured respectively along the X_1, X_2, and X_3 directions of the crystal are

$$\lambda_1 = \frac{2L}{n_1}; \quad \lambda_2 = \frac{2L}{n_2}; \quad \lambda_3 = \frac{2L}{n_3} \tag{4.29}$$

Employing the usual relationship between wave number k and wavelength λ we may write instead of 4.29

$$k_1 = \frac{\pi n_1}{L}; \quad k_2 = \frac{\pi n_2}{L}; \quad k_3 = \frac{\pi n_3}{L} \tag{4.30}$$

and if we let $k^2 = k_1^2 + k_2^2 + k_3^2$, it follows from 4.27 and 4.28 that

$$\omega_l = v_l k \tag{4.31}$$

and

$$\omega_t = v_t k \tag{4.32}$$

The standing waves 4.25 and 4.26 may be viewed as sums of sets of travelling waves described by the familiar function

$$A \exp[i(\boldsymbol{k} \cdot \boldsymbol{r} - \omega t)]$$

where $\boldsymbol{k}$ may now be identified as the wave vector with components k_1, k_2, and k_3 and $\boldsymbol{r}$ is a position vector.

The components k_1, k_2 and k_3 may be represented in a three-dimensional coordinate system which defines a $\boldsymbol{k}$ space (see Fig. 4.2); the points defined by eqn. 4.30 are then distributed as lattice points in the positive octant. The volume of the octant of radius k is

$$\tfrac{1}{6}\pi k^3$$

and the total number of lattice points in the octant is

$$\frac{Vk^3}{6\pi^2}$$

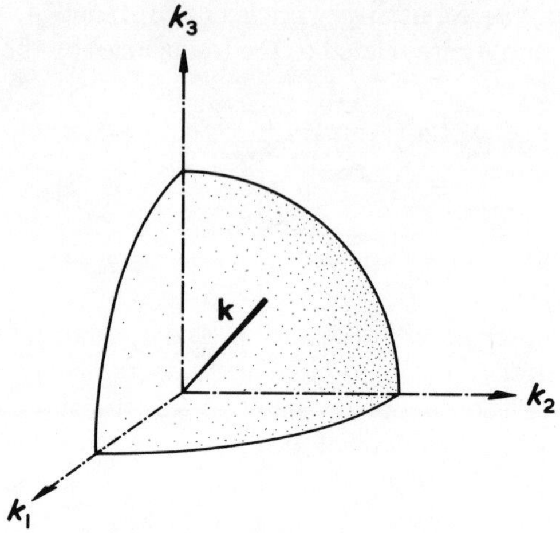

FIG. 4.2 Representation of wave vectors in **k**-space.

where $V(=L^3)$ is the volume of the crystal. For a particular frequency ω, substitution in eqns 4.31 and 4.32 will define two wave numbers k_l and k_t

$$k_l = \frac{\omega}{v_l} \tag{4.33}$$

and

$$k_t = \frac{\omega}{v_t} \tag{4.34}$$

Thus, the total number of lattice points in the octants of radii k_l and k_t are respectively

$$\frac{V}{6\pi^2}\left(\frac{\omega}{v_l}\right)^3 \quad \text{and} \quad \frac{V}{6\pi^2}\left(\frac{\omega}{v_t}\right)^3$$

But since with every lattice point, we may associate one longitudinal mode and two transverse modes, the total number of modes $M(\omega)$ of frequency less than ω is given by

$$M(\omega) = \frac{V}{6\pi^2}\left(\frac{1}{v_l^3} + \frac{2}{v_t^3}\right)\omega^3 \tag{4.35}$$

A mean velocity v may be defined by the equation

$$\frac{3}{v^3} = \frac{1}{v_l^3} + \frac{2}{v_t^3} \tag{4.36}$$

so that substitution of 4.36 in 4.35 leads to

$$M(\omega) = \frac{V}{2\pi^2 v^3}\,\omega^3 \tag{4.37}$$

Thus by differentiation of eqn. 4.37, we may obtain our required frequency distribution function $\mathcal{N}(\omega)$.

$$\mathcal{N}(\omega) = \frac{3V\omega^2}{2\pi^2 v^3} \tag{4.38}$$

This frequency distribution function was employed by Debye who also assumed that for a solid containing N atoms; the maximum frequency ω_m was determined by setting eqn. 4.37 equal to $3N$, which leads to

$$\frac{V}{v^3} = \frac{6N\pi^2}{{\omega_m}^3} \tag{4.39}$$

Returning now to eqn. 4.12 we may substitute therein by means of eqns 4.38 and 4.39 leading to

$$U = \frac{9N\hbar}{{\omega_m}^3}\int_0^{\omega_m} \frac{\omega^3\,\mathrm{d}\omega}{\exp(\hbar\omega/k_B T) - 1} \tag{4.40}$$

To simplify 4.40 we may let $x = \hbar\omega/k_B T$ so that

$$\mathrm{d}\omega = k_B\frac{T}{\hbar}\,\mathrm{d}x$$

Further if we let $\hbar\omega_m/k_B = \theta_D$, eqn. 4.40 becomes

$$U = 9Nk_B T\left(\frac{T}{\theta_D}\right)^3 \int_0^{\theta_D/T} \frac{x^3\,\mathrm{d}x}{e^x - 1} \tag{4.41}$$

Having laboured hard and long to obtain an expression for the internal energy, it is satisfying to note that at high temperature ($T \gg \theta_D$), since the integrand may be replaced by x^2, a limiting value of $C_v = 3Nk_B$ is obtained. At very low temperatures ($T \ll \theta_D$) the upper limit of the integral may be replaced by ∞.

Now since

$$\int_0^{\infty} \frac{x^3\,\mathrm{d}x}{e^x - 1} = \frac{\pi^4}{15}$$

the low temperature limit for the internal energy is given by the expression

$$U = \frac{3}{5}\pi^4 N k_B T\left(\frac{T}{\theta_D}\right)^3 \tag{4.42}$$

Hence

$$C_v = \frac{12}{5}\pi^4 N k_B \left(\frac{T}{\theta_D}\right)^3 \tag{4.43}$$

which is an expression of the famous Debye T^3 law for specific heats at low temperature. The general equation for C_v is also easily obtained

$$C_v = 9Nk_B\left(\frac{T}{\theta_D}\right)^3 \int_0^{\theta_D/T} \frac{e^x x^4}{(e^x - 1)^2}\,dx \tag{4.44}$$

but can only be integrated numerically. As an example of the satisfactory agreement found between theory and experiment for many materials, Fig. 4.3 shows data for graphite. Finally in Fig. 4.4 we compare the Einstein and

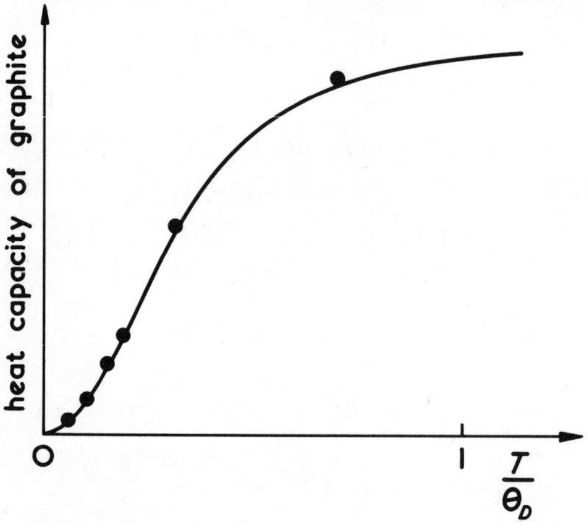

FIG. 4.3 Comparison of the Debye specific heat curve and the observed specific heat of graphite.

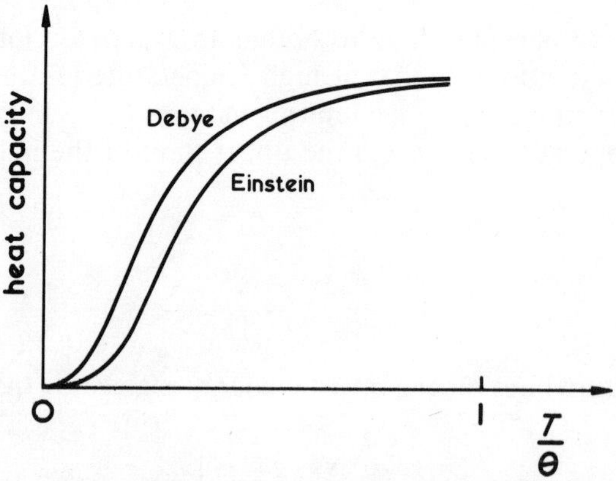

FIG. 4.4 Comparison of the Einstein and Debye specific heat equations.

Debye specific heat curves. The reader is left to decide the reason for the relative position of the curves at low temperature.

4.3 Elastic waves in one-dimensional atomic lattices

As must have become clear, the least satisfactory aspect of the Debye theory is the assumption that the frequency distribution appropriate to long-wavelength elastic modes in a continuum is appropriate for waves of shorter wavelength. It would be more satisfactory if the modes of vibration of the solid could be computed on a more rigorous basis. Born and von Kármán[3] developed a method for doing this, but generally the contemporary Debye theory tended to be regarded more favourably. However, much more recently a number of people have looked again more critically at the question of the frequency distribution, successfully employing an approach akin to that of Born and von Kármán. Our next task, therefore, is to consider vibrational modes of atomic lattices. It is fortunate that the major aspects of this topic emerge from an elementary discussion limited to one dimension.

The monatomic lattice

Let us consider initially an infinite one-dimensional array of point atoms of mass m at an equilibrium separation a (see Fig. 4.5). We will assume that

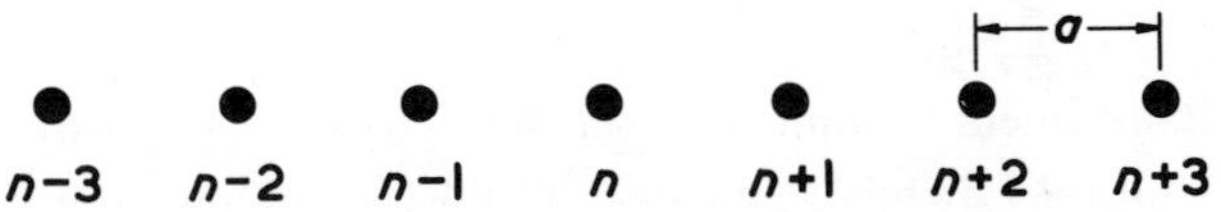

FIG. 4.5 The one-dimensional monatomic lattice.

there is a Hook's law interaction between nearest neighbours only. The atoms are labelled 0, 1, 2... $(n-1)$, n, $(n+1)$... and their respective displacements are $x_0, x_1, x_2 \ldots$ The equation of motion of the particle n is

$$m\frac{d^2x_n}{dt^2} = -f(x_n - x_{n-1}) - f(x_n - x_{n+1}) \tag{4.45}$$

$$= -f(x_{n-1} + x_{n+1} - 2x_n) \tag{4.46}$$

where f is the nearest neighbour force constant.

As is common practice in solving second order differential equations, let us assume a plane wave solution:

$$x_n(t) = \exp[i(kna - \omega t)] \tag{4.47}$$

where k is a wave number $= 2\pi/\lambda$ and na is the equilibrium position of atom

n relative to the origin. On substitution in 4.46 we obtain

$$m\omega^2 = -f[\exp(ika) + \exp(-ika) - 2]$$
$$= 4f\sin^2\left(\frac{ka}{2}\right) \tag{4.48}$$

or

$$\omega = \omega_{max}\sin\left(\frac{ka}{2}\right) \tag{4.49}$$

where

$$\omega_{max} = 2\left(\frac{f}{m}\right)^{1/2}$$

We have thus obtained an expression for the frequency of the waves in terms of the wave number k. Such a relationship is always referred to as a *dispersion relationship* and we assume the convention that to the wave number k, there corresponds a frequency $\omega(k)$. The dispersion relationship is illustrated in Fig. 4.6. We note immediately that the solution is periodic such that

$$\omega(k) = \omega\left(k + \frac{2\pi n}{a}\right) \tag{4.50}$$

where $n = 0, \pm 1, \pm 2$ etc.

Thus it is convenient to limit solutions for ω to a range of wave numbers equal to $2\pi/a$; it is normal to define for this purpose a *zone* such that

$$-\pi/a \leq k \leq \pi/a \tag{4.51}$$

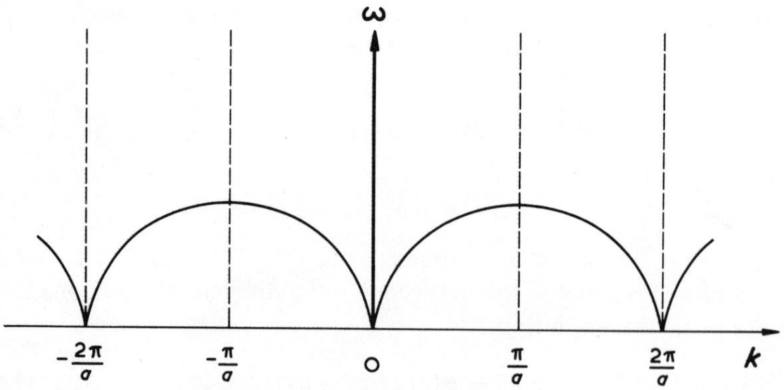

FIG. 4.6 The dispersion relationship for an infinite one-dimensional monatomic solid.

We note that since $2\pi/a$ is the shortest reciprocal lattice number of our linear lattice, eqn. 4.51 limits k to the *first Brillouin zone*. It is of interest to look a little more closely at the condition $k = \pm\pi/a$ (what we will shortly call the zone boundary). It may be assumed at this stage or for those familiar perhaps with physical optics it may be recalled that the velocity v_g of energy transmission in a medium, the group velocity of a wave packet is given by the equation

$$v_g = \frac{d\omega}{dk}$$

Therefore it will be observed that at the zero boundary $v_g = 0$; in other words at $k = \pm\pi/a$ our solution is no longer a travelling wave. We will refer again to the consequences of this observation in a somewhat different context.

If we wish to find an expression for the frequency distribution function $\mathcal{N}(\omega)$ for our linear array it is necessary to choose boundary conditions. We could make our crystal of finite length and assume that the terminal atoms were stationary, but we prefer to introduce instead *periodic boundary conditions* (they may be alternatively called cyclic or Born–von Kármán boundary conditions) which retain travelling wave solutions. We choose an unbounded lattice, but require that the displacement of an atom is periodic over a large distance $L = Na$ compared with a such that

$$x_n^{(t)} = x_{(n+N)}^{(t)} \tag{4.52}$$

(The reason for alternatively labelling these boundary conditions *cyclic* should be clear after a little thought.)

Applying condition 4.52 to our running wave solution 4.47, we find that

$$\begin{aligned} \exp(ikna) &= \exp[ik(n+N)a] \\ &= \exp(ikna)\exp(ikNa) \end{aligned} \tag{4.53}$$

or

$$\exp(ikL) = 1 \tag{4.54}$$

Hence according to 4.54, the wave numbers k are limited to

$$k = 0, \pm\frac{2\pi}{L}, \pm\frac{4\pi}{L}, \ldots, \pm\frac{N\pi}{L} \tag{4.55}$$

(if we limit k to the first Brillouin zone)

As an interesting extension we leave the reader to obtain an expression for $\mathcal{N}(\omega)$ and the keen reader to calculate the specific heat of the linear lattice. The *really* keen reader might then like to tackle the problem of $\mathcal{N}(\omega)$ in three dimensions (see Kittel, for example).

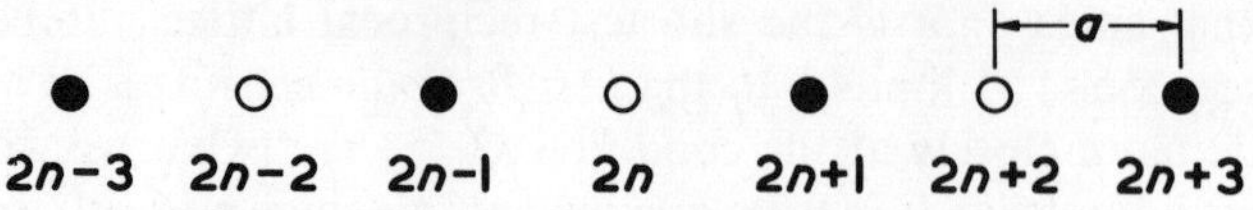

FIG. 4.7 The one-dimensional diatomic lattice.

The diatomic lattice

We consider the infinite diatomic lattice of Fig. 4.7, labelled so that the atoms designated even have mass m_1 and those designated odd have mass m_2. The treatment is entirely analogous to that used in the monatomic case. Therefore the equations of motion will be

$$m_1 \frac{d^2x_{2n}}{dt^2} = f(x_{2n-1} + x_{2n+1} - 2x_n) \tag{4.56}$$

and

$$m_2 \frac{d^2x_{2n+1}}{dt^2} = f(x_{2n} + x_{2n+2} - 2x_{2n+1}) \tag{4.57}$$

Again we may assume travelling waves of the type

$$x_{2n}(t) = A \exp[i(kna - \omega t)] \tag{4.58}$$

and

$$x_{2n+1}(t) = B \exp[i(kna - \omega t)] \tag{4.59}$$

Substitution in 4.56 and 4.57 yields the equations

$$(m_1\omega^2 - 2f)A + 2Bf \cos ka = 0 \tag{4.60}$$

and

$$(m_2\omega^2 - 2f)B + 2Af \cos ka = 0 \tag{4.61}$$

Non-trivial solutions for A and B exist only if the determinant of their coefficients is zero. i.e.

$$\begin{vmatrix} (m_1\omega^2 - 2f) & 2f \cos ka \\ 2f \cos ka & (m_2\omega^2 - 2f) \end{vmatrix} = 0 \tag{4.62}$$

Hence

$$\omega^2 = \frac{f}{\mu} \pm f\left[\frac{1}{\mu^2} - \frac{4\sin^2 ka}{m_1 m_2}\right]^{1/2} \tag{4.63}$$

where μ is the usual reduced mass.

Since ω must be positive, each value of ω^2 leads only to a single value of ω. But owing to the positive or negative terms in the right-hand side of

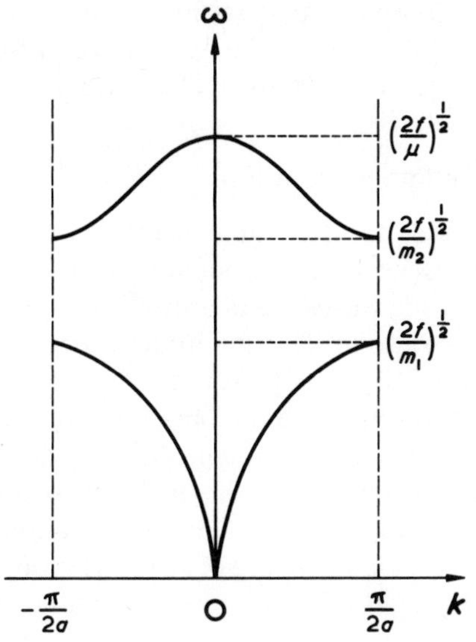

FIG. 4.8 The dispersion relationship for an infinite one-dimensional diatomic solid ($m_1 > m_2$).

eqn. 4.63 to each value of the wave number k there corresponds two values of ω. These are referred to as *branches*. Restricting our representation to the first Brillouin zone (see Fig. 4.8) we label the lower curve the *accoustic branch* and the upper curve the *optical branch*.

It is of value to enquire into the physical significance of the existence of the two branches. Some idea of this may be obtained by looking at the amplitude of vibrations of very small wave number k. In the limit of $k = 0$, it may be calculated that for the accoustical branch $A = B$ and for the optical branch $A/B = -m_2/m_1$. This implies that in the accoustical branch the two types of atoms may be considered as vibrating in phase whereas in the optical branch they vibrate out of phase, keeping their centre of mass fixed. The relationship between A and B for other values of k within the first Brillouin zone may be readily calculated by the reader.

If the atoms are electrically charged (i.e. an ionic crystal) the vibrations of the optical branch will be associated with an oscillating electric dipole in the crystal which can interact with the electric component of electromagnetic radiation. Indeed, infrared absorption is good evidence that sodium chloride *really* contains Na^+ and Cl^- ions and the reader might even consider the task of making the intensity of the infrared absorption of the sodium chloride crystal the bases of an estimate of the charges on the constituent ions.

Inspection of Fig. 4.8 reveals that solutions do not exist for frequencies between $(2f/m_1)^{1/2}$ and $(2f/m_2)^{1/2}$. The existence of a frequency gap at the boundary $k = \pm\, \pi/2a$ is a characteristic of a diatomic lattice.

4.4 Quantisation of lattice vibrations

Although we have tended to use a language which presumes a continuous spectrum of frequencies, nevertheless our treatment has demonstrated that lattice vibrations or elastic waves are quantised. This was the great triumph of the Einstein model of a solid. The quantum of energy in an elastic wave is the *phonon*, which is formally analogous to the familiar *photon*; the linguistic association of electromagnetic waves and photons is carried over to phonons and lattice vibrations. Evidence for quantisation perhaps more compelling than our calculations comes from the inelastic scattering of neutrons, during which process creation or annihilation of phonons occurs.

In our discussion of *elastic* scattering of X-ray photons by a crystal lattice, we arrived at what we might call a wave vector conservation law

$$\boldsymbol{k}_1 = \boldsymbol{k}_0 + \boldsymbol{G} \tag{4.64}$$

A photon of frequency ω_0, interacting with a lattice phonon, would carry a momentum

$$\boldsymbol{p}_0 = \hbar \boldsymbol{k}_0 \tag{4.65}$$

After the *inelastic* scattering process, its frequency would be changed to ω_1 and its momentum to $\hbar\boldsymbol{k}_1$. If in the process of scattering, a phonon of frequency ω_2 and wave vector $\boldsymbol{k}_2$ is created, two conservation equations are immediately suggested

(*i*) Conservation of energy:

$$\omega_0 = \omega_1 + \omega_2 \tag{4.66}$$

(*ii*) Conservation of wave vector:

$$\boldsymbol{k}_0 = \boldsymbol{k}_1 + \boldsymbol{k}_2 + \boldsymbol{G} \tag{4.67}$$

It will be noted that for completeness, Bragg reflexion has been included in the process. Second, it should be noted that although formally phonon momentum is implied in eqn. 4.67 (often referred to as crystal momentum) a phonon in a lattice does not really have momentum (see Kittel).

These conservation considerations constitute the basis of the very important experimental technique of inelastic neutron scattering, which can in principle give direct information about the phonon spectrum. The neutrons interact with the crystal nuclei and the scattering kinematics may be described by the conservation laws.

An example of inelastic scattering (for which the same conservation laws apply) familiar to chemists is Raman scattering, which involves the scattering of photons by optical phonons. The corresponding process involving acoustical phonons is known as Brillouin scattering.

References

(1) A. Einstein, *Ann. Physik* **22**, 180 (1907); **34**, 170 (1911)
(2) P. Debye, *Ann. Physik* **39**, 789 (1912)
(3) M. Born and Th. von Kârmân, *Physik Z.* **13**, 297 (1912); **14**, 15 (1913)

General

M. Born and K. Huang, *The Dynamical Theory of Crystal Lattices* (Clarendon Press, 1954)
W. Cochran, *The Dynamics of Atoms in Crystals* (Edward Arnold, 1973)

5

The Free Electron Gas

So far in this book we have limited our attention to those aspects of the solid state which may be discussed in terms of a model which assumes that the atomic constituents of a solid behave as neutral or charged spheres. We have neglected any discussion which might have involved more detailed assumption of atomic structure. For example, no general topic which we have discussed would really help in beginning to understand most of the characteristic properties of metals which are intimately associated with the behaviour of electrons in a lattice.

In this chapter we will begin to deal in an elementary way with the fascinating topic of electrons in metals by means of the *free electron theory*, the background to which is worthy of brief analysis.

Although the physical properties of metals had been extensively studied before 1900 (for example, the Wiedemann and Franz empirical law relating electrical and thermal conductivity was enunciated in 1853) real progress in understanding these properties quite reasonably did not emerge until after J. J. Thompson's discovery of the electron in 1897. Drude's crude theory of 1900, which was based on the assumption that metals contain a free electron gas, led surprisingly to a value of the Lorenz number in excellent agreement with experiment. However, separate formulae for electrical and thermal conductivities were unsatisfactory and even a thorough analysis of electron statistics based on the Boltzmann distribution by Lorentz in 1905 could not reconcile the physical behaviour of metals with a classical electron gas. The Drude–Lorentz theory was applied by numerous workers in the subsequent twenty years but it was clear that the failures outnumbered the successes. The first major success was achieved in 1927 when Pauli applied quantum statistics to explain the weak paramagnetism of the alkali metals, but major re-examination of the transport properties of metals was left to Sommerfeld[1] who published in 1928 his electron theory of metals based on Fermi–Dirac statistics. Although the Sommerfeld theory may be described as primitive, it is nevertheless worthwhile spending time in its development and examination; a significant insight into the properties of electrons in solids may be achieved.

The Sommerfeld or free electron theory of metals is perhaps better described as the statistical thermodynamical behaviour of a gas obeying Fermi–Dirac statistics. In concept it has very little to do with the solid state, since we adopt the very simple model of an electron gas in a container of large (relative to atomic) dimensions. In order to describe quantitatively the energy levels of an electron gas, we need to use some very simple quantum mechanical formulae and in order to describe the occupation of these levels, we need to know the appropriate distribution function. We will return to the latter point shortly but first we must focus our attention on elementary quantum mechanics. Our excursion into formal quantum mechanics will be painless and extend little further than the application of one or two basic principles. Although we will not luxuriate in the beauty of the discipline, for chemists without natural predilection towards quantum mechanics, or who are not too familiar with the postulatory approach, let us remind ourselves of some important aspects. In Chapter 6 we will deal separately with simple perturbation theory.

5.1 Postulates and working principles of quantum mechanics

Hamiltonian and quantum mechanics

It is a common practice to attempt to 'justify' the familiar Schrödinger equation by allusion to classical wave behaviour. This is probably a mistake, since the natural justification of quantum mechanics is that it works! Let us consider this point for a moment. The formulation of classical mechanics as embodied in the equations of Newton, Hamilton or Lagrange *is not self-evident*, but is accepted because it correctly predicts the behaviour of macroscopic bodies. Likewise we justify the *postulates* of quantum mechanics because they correctly predict the behaviour of sub-atomic bodies. However, we may see a *parallel relationship* between classical and quantum mechanics which is conceptually useful but a justification of neither. Let us see one aspect of this parallel relationship by inspecting the classical Hamiltonian.

The classical Hamiltonian function is an expression of the total energy (E) of a body in terms of its position ($\boldsymbol{r}$) and momentum ($\boldsymbol{p}$). For linear motion we may write

$$H(x, p) = \frac{p^2}{2m} + V(x) \tag{5.1}$$

Here we assume a *conservative* system and it may be simply shown (a useful exercise) from eqn. 5.1 that H may be alternatively related to the *canonically conjugate* variables p and x by the equations of motion:

$$\frac{-\partial H}{\partial x} = \dot{p} \tag{5.2}$$

and

$$\frac{\partial H}{\partial p} = \dot{x} \tag{5.3}$$

We may use the parallel relationship, if we replace the linear momentum p by the *linear operator*

$$\hat{p} = -i\hbar \frac{\mathrm{d}}{\mathrm{d}x} \tag{5.4}$$

(here we temporarily use to indicate an operator; no confusion should result from having used previously the identical symbol for a unit vector. Later we will adopt the normal practice of using the same symbol for both the operator and the dynamic variable). Instead of 5.1 we thus obtain the operator equation

$$\hat{H}(x, \hat{p}) = -\frac{\hbar^2}{2m}\frac{\mathrm{d}^2}{\mathrm{d}x^2} + V(x) \tag{5.5}$$

If we operate on some unindentified *wave function* $\psi(x)$ we obtain

$$\hat{H}(x, \hat{p})\psi = E\psi \tag{5.6}$$

and

$$-\frac{\hbar^2}{2m}\frac{\mathrm{d}^2\psi}{\mathrm{d}x^2} + V(x)\psi = E\psi \tag{5.7}$$

where solutions of 5.6 are known as eigenfunctions and the corresponding values of E are known as eigenvalues. Everyone will recognise 5.7 as the familiar one-dimensional time-independent Schrödinger equation. The connection between the classical Hamiltonian and the Schrödinger equation has been seen to result from the replacement of p by $\hat{p}$. This may be generalised by stating that *all* classical *dynamical variables* may be replaced by quantum mechanical *operators*, the statement constituting one of our basic *postulates* (see later.) But let us first assume an obvious extension into three-dimensional motion and write

$$\hat{H}(\mathbf{r}, \hat{p})\psi = E\psi \tag{5.8}$$

Now since classically we may write

$$H = \frac{1}{2m}(p_1{}^2 + p_2{}^2 + p_3{}^2) + V(\mathbf{r}) \tag{5.9}$$

where p_1, p_2, and p_3 are the momentum components; if we assume that

$$\hat{p}_1 = -i\hbar\frac{\partial}{\partial x_1} \tag{5.10}$$

$$\hat{p}_2 = -i\hbar\frac{\partial}{\partial x_2} \tag{5.11}$$

and

$$\hat{p}_3 = -i\hbar\frac{\partial}{\partial x_3} \tag{5.12}$$

we obtain the three-dimensional Schrödinger equation

$$-\frac{\hbar^2}{2m}\left(\frac{\partial^2\psi}{\partial {x_1}^2}+\frac{\partial^2\psi}{\partial {x_2}^2}+\frac{\partial^2\psi}{\partial {x_3}^2}\right)+V(\mathbf{r})\psi = E\psi \tag{5.13}$$

or writing

$$\nabla^2 = \frac{\partial^2}{\partial {x_1}^2}+\frac{\partial^2}{\partial {x_2}^2}+\frac{\partial^2}{\partial {x_3}^2}$$

$$\left[-\frac{\hbar^2}{2m}\nabla^2 + V(\mathbf{r})\right]\psi = E\psi \tag{5.14}$$

Quantum mechanical postulates

Having gone to some length to emphasise the potulatory nature of quantum mechanics let us now enumerate one possible set of postulates; we will not consider the question of *necessity* or *sufficiency*, although naturally the curious reader will wish to enquire further about this question.

(1) The state of a system can be represented by a function ψ. But for the function ψ to be acceptable, it must be single valued, finite and continuous and its first differential must be continuous. If these criteria are satisfied, ψ is known as a well-behaved function

(2) A further restriction is that the space integral

$$\int_{\text{space}} \psi^*\psi \, \mathrm{d}\tau$$

(where ψ^* is the complex conjugate of ψ) should be finite and positive and we note that where the integral is equal to unity, ψ is said to be normalised.

(3) Associated with every dynamical variable, there is a linear *Hermitian* operator (see Table 5.1).

(4) A measurement of a dynamic variable gives a result which is one of the eigenvalues of the corresponding operator.

TABLE 5.1 Linear operators

Dynamical variable	*Operator*
Position:	
x_1, x_2, x_3	x_1, x_2, x_3
Momentum:	
p_1, p_2, p_3	$-i\hbar\frac{\partial}{\partial x_1}, -i\hbar\frac{\partial}{\partial x_2}, -i\hbar\frac{\partial}{\partial x_3}$
Angular momentum:	
$l_1 = x_2p_3 - x_3p_2$	$-i\hbar\left(x_2\frac{\partial}{\partial x_3} - x_3\frac{\partial}{\partial x_2}\right)$
$l_2 = x_3p_1 - x_1p_3$	$-i\hbar\left(x_3\frac{\partial}{\partial x_1} - x_1\frac{\partial}{\partial x_3}\right)$
$l_3 = x_1p_2 - x_2p_1$	$-i\hbar\left(x_1\frac{\partial}{\partial x_2} - x_2\frac{\partial}{\partial x_1}\right)$
Hamiltonian:	
$\frac{1}{2m}(p_1{}^2 + p_2{}^2 + p_3{}^2) + V(\mathbf{r})$	$-\frac{\hbar^2}{2m}\nabla^2 + V(\mathbf{r})$

(5) The average (or expectation) value of a dynamical variable α for a system in a state ψ is given by

$$\langle\alpha\rangle = \frac{\int \psi^*\hat{\alpha}\psi\, \mathrm{d}\tau}{\int \psi^*\psi\, \mathrm{d}\tau}$$

Equipped with these postulates, let us see how far we can proceed in solving a very simple problem: that of a free particle moving in one dimension. The time-independent Schrödinger equation will be

$$-\frac{\hbar^2}{2m}\frac{\mathrm{d}^2\psi}{\mathrm{d}x^2} = E\psi \tag{5.15}$$

a solution of which will be the travelling wave

$$\psi = A\exp(ikx) \tag{5.16}$$

(where A is an arbitrary constant) leading to

$$E = \frac{\hbar^2k^2}{2m}$$

Inspection of postulates (1) and (2) will immediately show, however, that the solution of eqn. 5.15 for the free particle is not acceptable. We solve this problem by introducing *boundary conditions* (see Schiff) such as enclosing the particle in a one-dimensional box.

A particle in rectangular potential well

Practically everyone will be familiar with 'the particle in a box' so for variety let us consider a situation somewhat more general than normal. We refer to a particle constrained to a one-dimensional motion and subjected to a potential $V(x)$ which has a value of V_0 in regions 1 and 2 and zero in region 3 (see Fig. 5.1). We shall confine our attention to the solutions describing the

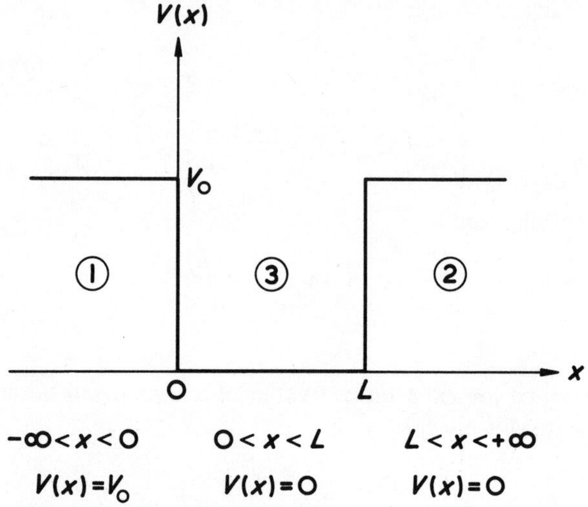

FIG. 5.1 An electron in a rectangular potential well.

bound states of energy $E < V_0$. We write the wave equations and solutions appropriate to the regions 1, 2 and 3.

In region 1:

$$-\frac{\hbar^2}{2m}\frac{d^2\psi}{dx^2} + V_0\psi = E\psi \tag{5.17}$$

with a general solution

$$\psi = A_1 \exp(\alpha x) + B_1 \exp(-\alpha x) \tag{5.18}$$

where A_1 and B_1 are arbitrary constants and

$$\alpha = \left(\frac{2m(V_0 - E)}{\hbar^2}\right)^{1/2}$$

In region 2: we have a wave equation identical with 5.17 with a general solution

$$\psi = A_2 \exp(\alpha x) + B_2 \exp(-\alpha x) \tag{5.19}$$

In region 3:

$$-\frac{\hbar^2}{2m}\frac{d^2\psi}{dx^2} = E\psi \tag{5.20}$$

with a general solution (written for convenience in real form)

$$= A_3 \cos \beta x + B_3 \sin \beta x \tag{5.21}$$

where

$$\beta = \left(\frac{2mE}{\hbar^2}\right)^{1/2}$$

Taking the positive root of the real quantity α, we note that in region 1, $B_1 \exp(-\alpha x) \to \infty$ as $x \to \infty$ so according to postulate (1), we must set $B_1 = 0$. A similar argument leads to $A_2 = 0$. Hence in the three regions we have

$$\psi = A_1 \exp(\alpha x) \tag{5.22}$$

$$\psi = B_2 \exp(-\alpha x) \tag{5.23}$$

$$\psi = A_3 \cos \beta x + B_3 \sin \beta x \tag{5.24}$$

We may evaluate the constants A_1, B_2 and A_3 by means of the continuity conditions [see postulate (1)]; this simple exercise is left to the reader who will confirm that

$$\tan \beta L = \frac{2\alpha\beta}{\beta^2 - \alpha^2} \tag{5.25}$$

Hence by substitution of the values of α and β we obtain

$$\tan\left[\left(\frac{2mE}{\hbar^2}\right)^{1/2} L\right] = \left[\frac{4E(V_0 - E)}{(2E - V_0)^2}\right]^{1/2} \tag{5.26}$$

Equation 5.26 may be solved numerically for E for any given values of L and V_0. But we see that the case of a particle in a box with infinite barriers is a special case and we may proceed by means of the simplifying assumption that $V_0 \to \infty$ at $x = 0$ and $x = L$, so that an analytical solution may be obtained. It should be confirmed that this new boundary condition leads to

$$\sin \beta L = 0 \tag{5.27}$$

or

$$\beta = \frac{n\pi}{L} \qquad \text{where} \quad n = 1, 2, 3 \ldots \tag{5.28}$$

Hence by substitution we may write for the eigenfunctions

$$\psi_n = B_3 \sin\left(\frac{n\pi x}{L}\right) \tag{5.29}$$

and for the eigenvalues

$$E_n = \frac{\hbar^2 \pi^2 n^2}{2mL^2} \qquad \text{where} \quad n = 1, 2, 3 \ldots \tag{5.30}$$

Normalisation of the wave functions may be effected in the usual way by setting

$$B_3{}^2 \int_0^L \sin^2\left(\frac{n\pi x}{L}\right) \mathrm{d}x = 1 \tag{5.31}$$

so that the *normalised* wave functions are

$$\psi_n = \left(\frac{2}{L}\right)^{1/2} \sin\left(\frac{n\pi x}{L}\right) \tag{5.32}$$

$$n = 1, 2, 3 \ldots$$

Although we have not discussed orthogonality of wave functions, those unfamiliar with the description should consult any book on elementary quantum mechanics and hence demonstrate that ψ_n constitute an orthonormal set in the whole range of x.

5.2 The free electron gas in one dimension at zero K

Having solved the wave equation for a *single electron* confined by a barrier of infinite potential, we consider whether or not we might use this result as the basis of a model for a one-dimensional metal. Physically we might envisage our hypothetical one-dimensional solid as a wire of length, say, 1 cm which would contain $\sim 10^8$ atoms. If we assume that each atom contributes *one* electron, we have to consider, therefore, 10^8 electrons confined to our wire by infinite barriers at each end. Neglecting the variation in potential due to the positive ion cores, we assume that the electrons move in a constant (for convenience, zero) potential. We emphasise that in our previous discussion, we were considering the dynamics of a single electron; now we have 10^8! Here and subsequently we will use the *one-electron approximation*, with which chemists should be very familiar. In this approximation, we assume, as would be the case if the Schrödinger equation were separable (think about this if it is not obvious), that each electron can be represented by its *own* wave function which is a function of the coordinates of *that electron* only. Thus the one-electron wave functions are calculated just as if the problem were that of a single electron moving in the average potential of the other electrons and that of any other source (such as the positive nuclei). Thus we may take unchanged eqns 5.30 and 5.32 to describe all of the electrons in our system (in common with many others we will use ε subsequently to identify specifically one-electron energy eigenvalues).

The elementary statement of the *Pauli exclusion principle* is that no two electrons can have an identical set of quantum numbers. But for any one-electron energy level labelled by a particular value of n, there are two possible spin orientations of the electron described by the spin quantum number $m_s = \pm\frac{1}{2}$. Hence in the familiar way we may accommodate our electrons at 0 K, two per energy level labelled by $n = 1, 2, 3 \ldots$, n extending as far as required for the total number of electrons.

We let n_0 denote the topmost filled energy level and, supposing for convenience that the total number of electrons N is an even number, it is evident that

$$2n_0 = N$$

We define the *Fermi energy* μ_0 as that of the topmost filled level i.e.

$$\mu_0 = \frac{\hbar^2\pi^2 n_o^{\,2}}{2mL^2} \tag{5.33}$$

$$= \frac{\hbar^2\pi^2 N^2}{8mL^2} \tag{5.34}$$

It is of interest to relate the average energy of the electrons $\langle\varepsilon\rangle$ to the Fermi energy. The total energy $\varepsilon_{\text{total}}$ may be calculated by summing twice the energies of each level from $n = 1$ to $n = n_0 = N/2$. Thus

$$\varepsilon_{\text{total}} = 2\sum_{n=1}^{N/2} \varepsilon_n \tag{5.35}$$

$$= \frac{\hbar^2\pi^2}{mL^2}\sum_{n=1}^{N/2} n^2 \tag{5.36}$$

Noting that for large p the summation

$$\sum_{n=1}^{p} n^2 \simeq \tfrac{1}{3}p^3$$

it follows from eqn. 5.34 that

$$\varepsilon_{\text{total}} = \frac{N\mu_0}{3} \tag{5.37}$$

or

$$\langle\varepsilon\rangle = \frac{\mu_0}{3} \tag{5.38}$$

Substitution of realistic numbers in the appropriate expressions for ε_n (this is a recommended exercise) will immediately indicate that the consecutive energy levels of the system are so close together that as in our discussion of phonons in Chapter 4 we may assume a continuum of *states* and define in the same way a *density of states* $\mathcal{N}(\varepsilon)$ i.e.

$$\mathcal{N}(\varepsilon) = 2\frac{dn}{d\varepsilon_n} \quad \text{(2 states/level)} \tag{5.39}$$

or from eqns 5.30 and 5.39

$$\mathcal{N}(\varepsilon) = \frac{2mL^2}{\hbar^2\pi^2 n} \tag{5.40}$$

$$= \left(\frac{2mL^2}{\hbar^2\pi^2\varepsilon_n}\right)^{1/2} \tag{5.41}$$

Using 5.41, we plot $\mathcal{N}(\varepsilon)$ as a function of energy in Fig. 5.2 and indicate filling of the states at 0 K by the shading up to the Fermi energy μ_0.

Having reached the stage of writing an expression for the density of one-electron states in one dimension, and before going on to begin a discussion of

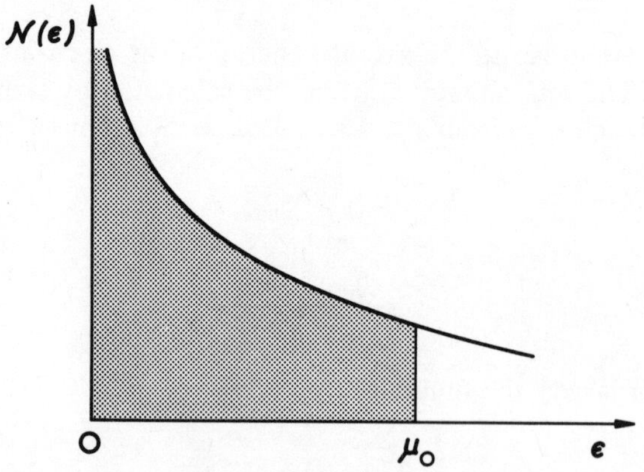

FIG. 5.2 Density of states in one dimension.

the filling of the states at finite temperatures, it will be useful for further reference to develop from a different viewpoint the energy of the electron states, but which will nevertheless lead to an identical expression for the density of states.

Earlier, we abandoned the unbounded electron wave function of eqn. 5.15 and instead, by use of simple boundary conditons, obtained the standing wave solutions of eqn. 5.32 for an electron confined within infinite barriers. This can be appreciated by examining the expectation value for the momentum $\langle p \rangle$ of the electron

$$\langle p \rangle = \int_0^L \psi_n^* \left(-i\hbar \frac{d\psi_n}{dx} \right) dx \tag{5.42}$$

Appropriate substitution and integration will show that

$$\langle p \rangle = 0 \tag{5.43}$$

which does not mean of course that the electron is stationary, but that at any instant the measurement of its momentum is equally likely to lead to a positive or negative value. Although standing waves are formally acceptable solutions, it is frequently more convenient to have travelling wave solutions (for example, in the examination of transport properties of free electrons in solids) of the type 5.16, which represents a particle with a constant momentum $\hbar k$. This is shown readily as follows:

$$-i\hbar \frac{d\psi}{dx} = p\psi(x) \tag{5.44}$$

(p is the eigenvalue of the momentum operator). Substitution from eqn. 5.16 leads immediately to

$$p = \hbar k \tag{5.45}$$

We note also that k is the wave-number $(=2\pi/\lambda)$.

Having emphasised the desirability of travelling wave solutions, how are they achieved? We recall from Chapter 4, that we utilised periodic boundary conditions in our discussion of the dispersion relationship. Again in the present context we may overcome the normalisation problem by the boundary condition

$$\psi(x) = \psi(x + L) \tag{5.46}$$

Hence from 5.16

$$\exp(ikx) = \exp[ik(x + L)] \tag{5.47}$$

Thus

$$\exp(ikL) = 1 \tag{5.48}$$

or

$$k = \frac{2\pi n}{L} \quad \text{where} \quad n = 0, \pm 1, \pm 2, \ldots \tag{5.49}$$

so that

$$\varepsilon_k = \frac{\hbar^2 \pi^2 (2n)^2}{2mL^2} \tag{5.50}$$

We note that we now permit both positive and negative values of n since $A \exp(ikx)$ and $A \exp(-ikx)$ represent different (although degenerate) states Normalisation leads to the wave function

$$\psi_k(x) = \left(\frac{1}{L}\right)^{1/2} \exp(ikx) \tag{5.51}$$

A comparison of eqns 5.30 and 5.50 indicates that half of the states which are included in the former are absent in the latter. However, whereas the energy levels ε_n have only a two-fold degeneracy, the energy levels ε_k have a four-fold degeneracy, which must mean that the density of one-electron states is identical. In other words, we must write instead of 5.39

$$\mathcal{N}(\varepsilon) = 4 \frac{\mathrm{d}n}{\mathrm{d}\varepsilon_k} \tag{5.52}$$

which leads to a final expression for $\mathcal{N}(\varepsilon)$ identical with that expressed in eqn. 5.41. Thus we may assume that since, as we will see, the bulk properties of our metal depend on the density of states function we may equivalently use standing or running wave solutions. Normally we will prefer the latter.

Having digressed somewhat in order to consider cyclic boundary conditions, we return to the problem of the filling of the one-electron states at a finite temperature. This is a common problem in elementary statistical mechanics so we must divert our attention temporarily to a brief introductory description of quantum statistics.

Introduction to quantum statistics

It is common practice in the elementary treatment of the statistical mechanics of a system of indistinguishable particles to justifiably assume that the number of eigenstates available to each particle is very much larger than the total number of particles in the system (N). Under these circumstances, we may utilise single particle functions (q) in order to define the ensemble partition function Q. Thus

$$Q = \frac{q^N}{N!} \tag{5.53}$$

$$q = \sum_j \omega_j \exp\left(-\frac{\varepsilon_j}{k_B T}\right) \tag{5.54}$$

where ε_j are the single particle energy eigenvalues
and ω_j is the usual degeneracy factor for the jth level.

It is implicit in the treatment of this classical (or Boltzmann) system that a state of the total system is defined by the quantum states of the individual particles. If, however, certain symmetry restrictions limit the wave functions describing the system, the particles cannot be independent, even if particle–particle interactions of a classical nature are absent. Indeed, single particle partition functions may no longer have any meaningful significance.

Let ψ be a wave function representing a state of our system of N indistinguishable particles; we may assume that when the coordinates of two particles are exchanged, since only the magnitude $\psi\psi^*$ must remain unchanged the sign of ψ may change or remain unchanged. According to this property, we may label ψ as *antisymmetrical* or *symmetrical* respectively and it is one of the basic tenets of quantum mechanics that there is a zero probability of transitions occurring between symmetric and antisymmetric states. It is sometimes said that each set of states thus constitutes an ergodic system. States of a system of particles of half-integral spin (e.g. electrons and protons) have antisymmetric wave functions while whose of particles with integral spin (e.g. photons) have symmetric wave functions. Similarly, states of a system containing particles made up from an odd number of electrons, protons and neutrons (e.g. He^3) are described by antisymmetric wave functions while those of particles made up from an even number (e.g. H_2, He^4) are described by symmetric wave functions.

Although the classical (high temperature) limit of Boltzmann statistics is acceptable when dealing with numerous systems of interest and importance (even as we will see in dealing sometimes with electrons in solids) it is frequently necessary to utilise *Bose–Einstein statistics* (symmetric eigenfunctions) and *Fermi–Dirac statistics* (antisymmetric eigenfunctions). Conceptually we may distinguish these two types of quantum statistics by stating that in the former, the number of particles in a quantum state is unlimited, whereas in the latter, a maximum of one particle only is allowed. Systems for which quantum statistics must be applied are called *degenerate*.

It would be unwise to attempt a simple derivation of the basic Bose–Einstein and Fermi–Dirac distribution functions, so let it be sufficient to quote them without justification as

$$f(\varepsilon_j) = \frac{1}{\exp\left(\alpha + \frac{\varepsilon_j}{k_B T}\right) \pm 1} \tag{5.55}$$

where $f(\varepsilon_j)$ is the probability of occupation of the state of energy ε_j. The positive sign in the denominator is appropriate for Fermi–Dirac statistics and the negative sign for Bose–Einstein statistics. The parameter α is determined by the condition that the total number of particles N is fixed so that

$$\sum_j f(\varepsilon_j) = N \tag{5.56}$$

at all temperatures and may be identified by the relationship

$$\alpha = -\frac{\mu}{k_B T} \tag{5.57}$$

where μ is the chemical potential (frequently referred to in solid state physics as the *Fermi level*). In our discussions of electron statistics we will be using the Fermi–Dirac function which we will write as

$$f(\varepsilon) = \frac{1}{\exp\left(\frac{\varepsilon - \mu}{k_B T}\right) + 1} \tag{5.58}$$

We thus have an expression which gives the probability that a state of energy ε will be occupied at equilibrium at the absolute temperature T. It is instructive to plot the Fermi–Dirac function as in Fig. 5.3 from which it will be

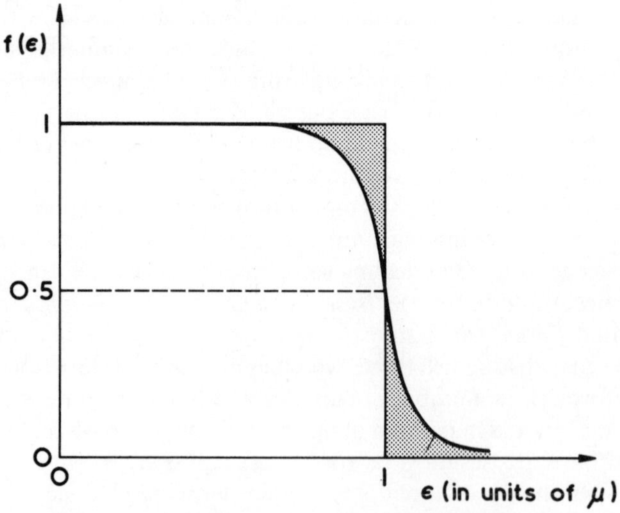

FIG. 5.3 The Fermi–Dirac distribution function.

observed that all finite temperatures $f(\varepsilon) = \frac{1}{2}$ when $\varepsilon = \mu$. Returning to our one-dimensional metal it is now quite obvious why we chose the symbol μ_0 for the Fermi energy; at 0 K, $\mu_0 = \mu$ because in the limit $T \to 0$, the function $f(\varepsilon)$ changes discontinuously from the value 1 to the value of 0 at $\varepsilon = \mu_0 = \mu$. At any finite temperature the density of occupied states will be given by

$$\mathcal{N}(\varepsilon) f(\varepsilon)$$

while μ will be determined by the equation

$$\int_0^\infty \mathcal{N}(\varepsilon) f(\varepsilon)\, d\varepsilon = N \tag{5.59}$$

Integrals of the type given in 5.59 will be commonly found in subsequent discussions (although with a subtle difference which should become apparent to the careful reader) and we will deal with a general solution in the next section when considering the three-dimensional electron gas.

5.3 The free electron gas in three dimensions

Having dealt in some detail with the one-dimensional electron gas and introduced Fermi–Dirac statistics it should now be possible to simply extend our discussion to the more realistic three-dimensional system (*ab uno disce omnes!*). The Schrödinger equation for a free electron in three-dimensions is

$$-\frac{\hbar^2}{2m}\nabla^2\psi_{\boldsymbol{k}} = \varepsilon\psi_{\boldsymbol{k}} \tag{5.60}$$

Again it is convenient to introduce periodic boundary conditions such that

$$\begin{aligned} \psi(x_1, x_2, x_3) &= \psi(x_1 + L, x_2, x_3) \\ &= \psi(x_1, x_2 + L, x_3) \\ &= \psi(x_1, x_2, x_3 + L) \end{aligned} \tag{5.61}$$

and the interested reader will readily confirm that the normalised wave functions are

$$\psi_{\boldsymbol{k}} = \left(\frac{1}{L}\right)^{3/2} \exp(i\boldsymbol{k}\cdot\boldsymbol{r}) \tag{5.62}$$

where the components of the position vector $\boldsymbol{r}$ are x_1, x_2 and x_3 and the components of the wave vector $\boldsymbol{k}$ have allowed values

$$k_1 = \frac{2\pi n_1}{L};\quad k_2 = \frac{2\pi n_2}{L};\quad k_3 = \frac{2\pi n_3}{L}$$

where n_1, n_2 and n_3 are 0, ± 1, $\pm 2\ldots$

The energy eigenvalues are evidently given by

$$\varepsilon_{\boldsymbol{k}} = \frac{\hbar^2 k^2}{2m} \tag{5.63}$$

$$= \frac{\hbar^2}{2m}(k_1{}^2 + k_2{}^2 + k_3{}^2) \tag{5.64}$$

$$= \frac{2\hbar^2\pi^2}{mL^2}(n_1{}^2 + n_2{}^2 + n_3{}^2) \tag{5.65}$$

The magnitude k of the wave vector is related to the wavelength λ by the usual equation

$$k = 2\pi/\lambda \tag{5.66}$$

and the linear momentum $\boldsymbol{p}$ is obviously related to the wave vector by the equation

$$\boldsymbol{p} = \hbar\boldsymbol{k} \tag{5.67}$$

The Fermi energy

Now let us suppose that we wish to accommodate N electrons at 0 K. According to the one-electron model we feed our N electrons progressively into the energy levels taking account of the Pauli principle. For our purposes we may assume that the quantum numbers describing each level are k_1, k_2 and k_3 (for one allowed wave vector $\boldsymbol{k}$) and the spin quantum number $(m_s) = \pm\frac{1}{2}$. A representation of this may be found conveniently by considering a lattice of points in $\boldsymbol{k}$ space (see Fig. 5.4), a treatment analogous to

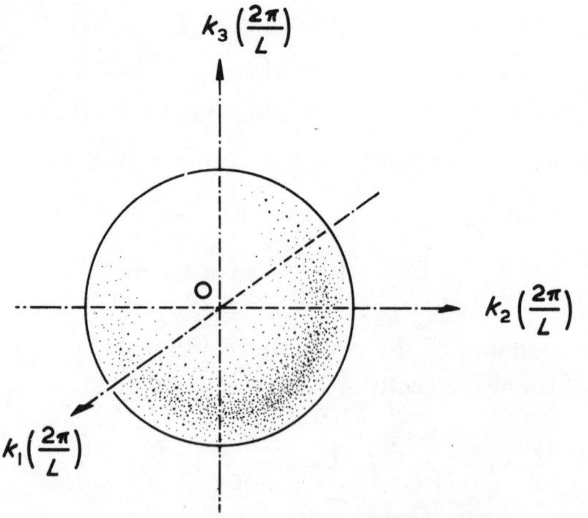

FIG. 5.4 Allowed quantum states represented in $\boldsymbol{k}$-space.

that used in Chapter 4 (Fig. 4.2). Every lattice point represents an allowed wave vector defined by its components k_1, k_2, and k_3 (in units of $2\pi/L$). Thus the occupation of electron states may be represented by a sphere in $\boldsymbol{k}$-space of radius k_0, the magnitude of the wave vector corresponding to the Fermi energy μ_0 (two electrons at each lattice point). The enclosed volume in $\boldsymbol{k}$ space ($\frac{4}{3}\pi k_0{}^3$) will thus contain $L^3 k_0{}^3/3\pi^2$ states, which must equal N, the number of electrons. Thus

$$k_0 = \left(\frac{3\pi^2 N}{L^3}\right)^{1/3} \tag{5.68}$$

Substitution in 5.63 leads to

$$\mu_0 = \frac{\hbar^2}{2m}\left(\frac{3\pi^2 N}{L^3}\right)^{2/3} \tag{5.69}$$

Thus we may take as our standard equation for the Fermi energy

$$\mu_0 = \frac{\hbar^2}{2m}\left(\frac{3\pi^2 N}{V}\right)^{2/3} \tag{5.70}$$

The density of states

Since we are discussing one-electron states, in any polyelectronic system the number of states in the interval ε and $\varepsilon + d\varepsilon$ must be the same as the number of electrons whose energy is included in this range. We know from a more general version of eqn. 5.70, that the energy of the qth electron is given by the equation

$$\varepsilon_q = \frac{\hbar^2}{2m}\left(\frac{3\pi^2 q}{V}\right)^{2/3} \tag{5.71}$$

so that on differentiation, the expression

$$\frac{dq}{d\varepsilon} = \frac{V}{2\pi^2}\left(\frac{2m}{\hbar^2}\right)^{3/2} \varepsilon_q^{1/2}$$

may be directly identified with the density of states function.
Thus

$$\mathcal{N}(\varepsilon) = \frac{V}{2\pi^2}\left(\frac{2m}{\hbar^2}\right)^{3/2} \varepsilon^{1/2} \tag{5.72}$$

It is sometimes convenient to state $\mathcal{N}(\varepsilon)$ in terms of μ_0 given by eqn. 5.70. Direct substitution leads to

$$\mathcal{N}(\varepsilon) = \frac{3N}{2\mu_0^{3/2}}\, \varepsilon^{1/2} \tag{5.73}$$

The total energy of the electron gas may be found, for variety, by a method different from that in section 5.2 by use of the density of states function:

$$\varepsilon_{\text{total}} = \int_0^{\mu_0} \varepsilon \mathcal{N}(\varepsilon)\, d\varepsilon \tag{5.74}$$

The use of 5.73 in the integral leads directly to

$$\varepsilon_{\text{total}} = \tfrac{3}{5} N \mu_o \tag{5.75}$$

or

$$\langle \varepsilon \rangle = \tfrac{3}{5} \mu_0 \tag{5.76}$$

Thermodynamic functions at zero K

The Helmholtz free energy F is given by the equation

$$F = U - TS$$

At zero K, therefore, we may write

$$F_0 = U_0 \tag{5.77}$$

We know further from classical thermodynamics that the pressure P is related to F by the equation

$$P = -\left(\frac{\partial F}{\partial V}\right)_T \tag{5.78}$$

Hence the zero point pressure P_0 may be expressed in terms of μ_0 by means of eqns 5.75, 5.77 and 5.78:

$$\begin{aligned} P_0 &= -\frac{\partial U_0}{\partial V} \\ &= -\frac{\partial \varepsilon_{\text{total}}}{\partial V} \\ &= -\frac{3}{5} N \frac{\partial \mu_0}{\partial V} \\ &= \frac{2N}{5V} \mu_0 \end{aligned} \tag{5.79}$$

By use of eqn. 5.70 we may make an appropriate estimate of μ_0 by assuming for simplicity that the metal supplies one electron/atom volume and hence by use of 5.79 calculate the expected zero point pressure. Abandoning SI units temporarily and preferring a more picturesque language, it should be confirmed that $P_0 \simeq 10^4$ atmospheres. Clearly an electron gas at such a high pressure requires a very strong container—the metal itself! One can think more constructively of the strength of the container being provided by the interaction between the electrons and the positive ion cores which results in the very low potential of the former with respect to the vacuum level.

Fermi–Dirac integrals

At temperatures other than zero K, thermodynamic functions of the electron gas may be readily calculated using the density of states function and the Fermi–Dirac distribution function. However, a series of related integrals, often referred to as Fermi–Dirac integrals, which cannot be solved in closed form and which are cumbersome to handle, are involved and it is useful to deal with them at the outset. Not that we will dwell on methods of solution, but merely quote the solutions and refer the mathematically inclined student to Mayer and Mayer. The general integral is:

$$I = \int_0^\infty F(\varepsilon) f(\varepsilon)\, \mathrm{d}\varepsilon \tag{5.80}$$

where $f(\varepsilon)$ is the Fermi–Dirac distribution function and
$F(\varepsilon)$ is any simple continuous function of ε such that
$F(\varepsilon) = 0$ at $\varepsilon = 0$.

Partial integration of I yields an interesting alternative form

$$I = G(\infty) f(\infty) - G(0) f(0) - \int_0^\infty G(\varepsilon) \frac{\mathrm{d}f(\varepsilon)}{\mathrm{d}\varepsilon}\, \mathrm{d}\varepsilon \tag{5.81}$$

where

$$G(\varepsilon) = \int_0^\varepsilon F(\varepsilon')\, \mathrm{d}\varepsilon'.$$

Simple arguments will show that the first two terms of eqn. 5.81 are both zero, so that

$$I = -\int_0^\infty G(\varepsilon) \frac{\mathrm{d}f(\varepsilon)}{\mathrm{d}\varepsilon}\, \mathrm{d}\varepsilon \tag{5.82}$$

The solution to 5.80 and 5.82 is expressed as follows:

$$I = \int_0^\mu F(\varepsilon)\, \mathrm{d}\varepsilon + \frac{\pi^2}{6}(k_{\mathrm{B}}T)^2 \left[\frac{\mathrm{d}F(\varepsilon)}{\mathrm{d}\varepsilon}\right]_{\varepsilon=\mu} + \frac{7\pi^4}{360}(k_{\mathrm{B}}T)^4 \left[\frac{\mathrm{d}^3F(\varepsilon)}{\mathrm{d}\varepsilon^2}\right]_{\varepsilon=\mu} + \cdots \tag{5.83}$$

As a simple example, we may develop an expression for the chemical potential of the electron gas by evaluating the integral analogous to 5.59

$$\int_0^\infty \mathcal{N}(\varepsilon) f(\varepsilon)\, d\varepsilon = N \tag{5.84}$$

which we observe is of the form 5.80, but may be written in a simpler form by use of 5.73.

$$N = \frac{3N}{2\mu_0^{3/2}} \int_0^\infty \varepsilon^{1/2} f(\varepsilon)\, d\varepsilon \tag{5.85}$$

We may simply write down the terms of 5.83 by appropriate substitution from 5.85.

$$\int_0^\mu \varepsilon^{1/2}\, d\varepsilon = \frac{2\mu^{3/2}}{3}$$

$$\left[\frac{d}{d\varepsilon}(\varepsilon^{1/2})\right]_{\varepsilon=\mu} = \frac{\mu^{-1/2}}{2}$$

$$\left[\frac{d^3}{d\varepsilon^3}(\varepsilon^{1/2})\right]_{\varepsilon=\mu} = \frac{3\mu^{-5/2}}{8}$$

Hence instead of 5.85 we may write

$$1 = \left(\frac{\mu}{\mu_0}\right)^{3/2} \left[1 + \frac{\pi^2}{8}\left(\frac{k_B T}{\mu}\right)^2 + \frac{7\pi^4}{640}\left(\frac{k_B T}{\mu}\right)^4 + \cdots\right] \tag{5.86}$$

or

$$\mu = \mu_0 \left[1 + \frac{\pi^2}{8}\left(\frac{k_B T}{\mu}\right)^2 + \frac{7\pi^4}{640}\left(\frac{k_B T}{\mu}\right)^4 + \cdots\right]^{-2/3} \tag{5.87}$$

We may expand the right-hand side of eqn. 5.87 by means of the binomial theorem as far as the quadratic term, neglecting powers beyond the quartic. This procedure leads to the expression

$$\mu = \mu_0 \left[1 - \frac{\pi^2}{12}\left(\frac{k_B T}{\mu}\right)^2 + \frac{\pi^4}{720}\left(\frac{k_B T}{\mu}\right)^4 + \cdots\right] \tag{5.88}$$

In order to effectively separate μ and μ_0 we use the approximation (obtained by assuming

$$\mu \simeq \mu_0 \left[1 + \frac{\pi^2}{6}\left(\frac{k_B T}{\mu}\right)^2\right]^{-1/2}$$

and substituting $\mu = \mu_0$):

$$\mu^{-2} = \mu_0^{-2}\left[1 + \frac{\pi^2}{6}\left(\frac{k_B T}{\mu_0}\right)^2\right]$$

which is substituted in the quadratic term of 5.88. In the quartic term of 5.88 the direct substitution $\mu = \mu_0$ is made. We thus obtain

$$\mu = \mu_0\left[1 - \frac{\pi^2}{12}\left(\frac{k_B T}{\mu_0}\right)^2 - \frac{\pi^4}{80}\left(\frac{k_B T}{\mu_0}\right)^4 \cdots\right] \tag{5.89}$$

This a very important equation relating the chemical potential to the Fermi energy; the serious reader will have cause to refer repeatedly to this equation.

Thermodynamic functions of the free electron gas

Having established the method of dealing with Fermi–Dirac integrals by reference to the calculation of the chemical potential it is now a straightforward matter to calculate other thermodynamic functions.

The internal energy U may be calculated from the equation

$$U = \int_0^\infty \varepsilon \mathcal{N}(\varepsilon) f(\varepsilon)\, d\varepsilon \tag{5.90}$$

or alternatively by use of eqn. 5.73

$$U = \frac{3N}{2\mu_0^{3/2}} \int_0^\infty \varepsilon^{3/2} f(\varepsilon)\, d\varepsilon \tag{5.91}$$

Clearly 5.91 is very similar to 5.85 and may be solved in an identical way. It would be good practice in the use of the solution of a Fermi–Dirac integral to confirm the following expression:

$$U = \frac{3}{5} N\mu\left(\frac{\mu}{\mu_0}\right)^{3/2}\left[1 + \frac{5\pi^2}{8}\left(\frac{k_B T}{\mu}\right)^2 - \frac{7\pi^4}{384}\left(\frac{k_B T}{\mu}\right)^4 + \cdots\right] \tag{5.92}$$

By use of 5.89 we may replace μ by μ_0 to give

$$U = \frac{3}{5} N\mu_0\left[1 + \frac{5\pi^2}{12}\left(\frac{k_B T}{\mu_0}\right)^2 - \frac{\pi^4}{16}\left(\frac{k_B T}{\mu_0}\right)^4 + \cdots\right] \tag{5.93}$$

(this is a somewhat tricky manipulative exercise)

The heat capacity C_v may be found by direct differentiation of 5.93:

$$C_v = \frac{dU}{dT} = \frac{Nk_B\pi^2}{2}\left(\frac{k_B T}{\mu_0}\right)\left[1 - \frac{3\pi^2}{10}\left(\frac{k_B T}{\mu_0}\right)^2 + \cdots\right] \tag{5.94}$$

From 5.94 we may obtain an expression for the entropy of the electron gas by use of the equation

$$S = \int_0^T \frac{C_v}{T}\,\mathrm{d}T$$

Hence

$$S = \frac{Nk_B\pi^2}{2}\left(\frac{k_B T}{\mu_0}\right)\left[1 - \frac{\pi^2}{10}\left(\frac{k_B T}{\mu_0}\right)^2 + \cdots\right] \tag{5.95}$$

$$= \frac{N\mu_0}{T}\left[\frac{\pi^2}{2}\left(\frac{k_B T}{\mu_0}\right)^2 - \frac{\pi^4}{20}\left(\frac{k_B T}{\mu_0}\right)^4 + \cdots\right] \tag{5.96}$$

For fear of tedium we will not continue! It is nevertheless not without interest to complete the collection of thermodynamic functions and it is hoped the reader will do it at leisure. He or she will surely be conscious that in all the thermodynamic properties of an electron gas, the temperature dependent part always occurs as $k_B T/\mu_0$, which is about $10^{-2} - 10^{-3}$ at ordinary temperatures. This makes the thermodynamic functions rather temperature insensitive.

5.4 Simple application of the free electron theory

The practically minded reader of this chapter may be critical, with some justification, in as far as we have offered no tangible experimental observations to consolidate our crude theory. We hope, at least in part, to rectify this situation by pointing out some ways in which the free electron theory may be used as the basic rationale of experimental data. In the first instance we make use of our discussion of thermodynamic functions to discuss heat capacities, magnetic properties, and (perhaps surprisingly) thermionic emission.

Heat capacities of metals

Calculations based on classical mechanics predict that a gas of N free particles should have a heat capacity of $3/2\ Nk_B$. But if it is assumed that N atoms in a solid supply a total of N free electrons, the contribution of the latter to the total heat capacity is $\sim 10^{-2}$ of this value. Of course, the apparent paradox, which troubled theoretical physicists such as Lorentz, was that while the electrons appeared to be free to participate in electrical conduction processes, they were not free to contribute to the heat capacity. We now know the answer. According to the Fermi–Dirac distribution function (see Fig. 5.3) at any finite temperature only a fraction (those within the range $\sim k_B T$ of $\varepsilon/\mu = 1$) of the total electron population may become thermally excited. We indicate this by the shaded regions in Fig. 5.3 where electrons at a finite temperature may be thought of as being transferred from the shaded region below to the shaded region above $\varepsilon/\mu = 1$. This qualitative description is stated formally in eqn. 5.94, where taking only the first term we predict a linear temperature dependence of the heat capacity of the electron gas. The difference between the temperature dependence of the heat capacity contributions of electrons and lattice vibrations at low temperatures immediately suggests a method of experimental separation by plotting C_v/T versus T^2. An example of this analysis is shown in Fig. 5.5 and a list of electronic heat capacities is given for example, in Kittel. Comparison of experimentally determined electronic heat capacities with those calculated from eqn. 5.94 illustrate a broad agreement which may be made exact by assigning an *effective mass* to the electrons different from the normal value. For example, the ratios of the *thermal effective mass* to the normal mass for potassium, rubidium and caesium are respectively 1·25, 1·26 and 1·43. We prefer not to attempt any justification of the procedure, but prefer to think of the thermal effective mass as a conveniently adjustable parameter.

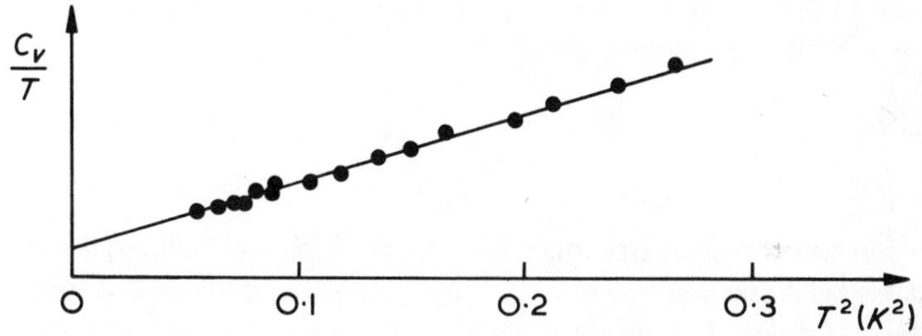

FIG. 5.5 The experimental heat capacity of potassium at low temperatures (after Lien and Phillips, *Phys. Rev.* **133A**, 1370 (1964)).

Thermionic emission

All metals (and incidentally many other interesting materials) emit electrons when heated to a sufficiently high temperature. This process, known as thermionic emission, may be regarded pictorially as evaporation of electrons from the surface, so it is reasonable that we should enquire the extent to which our free electron model can describe the main features of the process. Since we are dealing with a surface behaviour, it is desirable that we should refer to a zero of potential outside the surface. But so far our thermodynamic functions have been calculated on the basis of a zero of potential inside the metal. It is appropriate, therefore, we should refer to the vacuum level outside the surface by assigning to our constant potential inside the surface a value of $-\phi_0$ per electron.

Since electrons do not leave the surface of the metal at 0 K it follows that $\phi_0 > \mu_0$. At higher temperatures, however, as the kinetic energy of the electrons increases, a fraction at any temperature will have sufficient energy to surmount the potential barrier and hence leave the surface. The calculation of the rate of emission of electrons per unit area from the surface may be made by the well-known kinetic analysis found in all books on solid state physics. However, we may use alternatively the following simple thermodynamic argument.

At any finite temperature, the metal will be in equilibrium with electron gas in the space surrounding it. We will assume that the electrical potential in the vicinity of the metal surface is zero (see later); we could imagine, for example, the presence of an equivalent concentration of a positive ion gas to balance the electronic charge. Under this simplifying condition, the pressure of the electron gas surrounding the metal may be determined by necessarily equating its chemical potential with that of electron gas inside the metal:

$$-\phi = \mu - \phi_0 \qquad (5.97)$$

ϕ is known as the work function of the metal, assumed to a good approximation, to be temperature insensitive and since it is large, we further assume $\phi/k_B T \gg 1$. In the expression

$$f(\varepsilon) = \frac{1}{\exp\left(\dfrac{\varepsilon + \phi}{k_B T}\right) + 1}$$

the unity in the denominator may be neglected, so that classical functions may be used. Hence we may simply write down from statistical mechanics the chemical potential of the electrons in terms of the pressure (not forgetting an electronic degeneracy of 2)

$$-\phi = k_B T \ln\left[\left(\frac{h^2}{2\pi m k_B T}\right)^{3/2} \frac{P}{2k_B T}\right] \qquad (5.98)$$

or

$$P = 2k_B T\left(\frac{2\pi m k_B T}{h^2}\right)^{3/2} \exp(-\phi/k_B T) \qquad (5.99)$$

Two further assumptions may be made. At equilibrium, the rate of emission of electrons from the surface is equal to the rate of uptake of electrons from the gas. This latter process may be described in classical kinetic terms: the number of particles colliding with unit area in unit time (Z) is related to the pressure by

$$Z = \frac{P}{(2\pi m k_B T)^{1/2}} \qquad (5.100)$$

Second it will be assumed for simplicity that the reflexion coefficient at the surface is zero i.e. there is no internal reflexion of electrons provided the energetic conditions for escape are satisfied (this cannot affect the equilibrium, only the quantity).

The current density (J) is then given by the equation

$$J = Ze \tag{5.101}$$

So combining 5.99–5.101 we obtain

$$J = \left(\frac{k_B T}{\pi}\right)^2 \frac{me}{2\hbar^3} \exp\left(-\frac{\phi}{k_B T}\right) \tag{5.102}$$

which is the Richardson–Dushman equation. The temperature dependence of J according to 5.102 is found to be satisfactorily obeyed in practice and leads to values of work function similar to but not identical with those obtained by alternative methods. Some values of thermionic work functions together with the temperature independent parts of the pre-exponential factor (A) are given in Table 5.2. According to the Richardson–Dushman equation A should be constant ($1{\cdot}2 \times 10^6 \text{Am}^{-2}\text{K}^{-2}$) but clear differences are observed, the simple reasons for which the reader is left to consider.

TABLE 5.2 Thermionic work functions

	Richardson constants	
Metal	$A(\text{Am}^{-2}\text{K}^{-2})$	ϕ(eV)
Cs	$16{\cdot}0 \times 10^5$	1·81
Ba	$6{\cdot}0 \times 10^5$	2·11
Ni	$6{\cdot}0 \times 10^5$	4·10
Ta	$6{\cdot}0 \times 10^5$	4·10
Mo	$5{\cdot}5 \times 10^5$	4·15
W	$8{\cdot}0 \times 10^5$	4·54

Although not strictly consistent with its title we conclude this section with a few further remarks about work functions, which should be of interest to chemists; a book on elementary physics of solids should contain a discussion of work functions and it is difficult to see where alternatively the topic could be considered. In our initial discussion of thermionic emission we represented the metal-vacuum level interface as a sharply defined discontinuity in the potential. In fact, this is unrealistic since the escaping electron is subject to the so-called *image force* which has its origin in the electrostatic interaction of the electronic charge with the charge of opposite sign induced at the metal surface. This situation is dealt with in classical electrostatics by a force between two point charges at the electron-electron image separation i.e.

$$F_{\text{image}} = \frac{e^2}{16\pi\varepsilon_0 x^2} \tag{5.103}$$

where x is the distance between the electron and the surface. Equation 5.103 leads to the following expression for the potential

$$\Phi_{\text{image}} = \frac{-e^2}{16\pi\varepsilon_0 x} \tag{5.104}$$

which is represented graphically in Fig. 5.6 (note the inappropriateness of the image potential as $x \to 0$). Since the work function ϕ is defined with respect to a zero which really represents the potential of the electron at

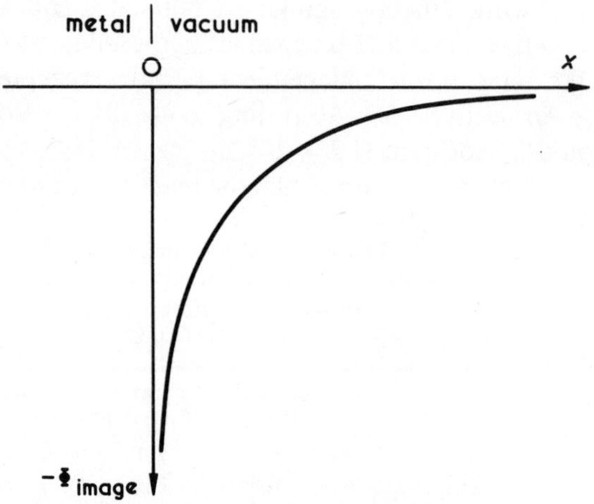

FIG. 5.6 The image force.

infinite distance from the metal surface, we might consider therefore that it is made up from two components: a component which is a consequence of the chemical potential of the electron and a component which is a consequence of the image force. It must be emphasised that besides representing the potential in the simplest way, the model is idealised in two other respects:

(1) There are no external fields present at the surface, and
(2) The surface is taken to be an infinite single crystal face represented by a uniformly conducting continuum.

Let us discuss these complications in turn. What will be the consequences of an accelerating electric potential at the surface? Inspection of Fig. 5.7 shows that the existence of the accelerating potential lowers the image barrier to

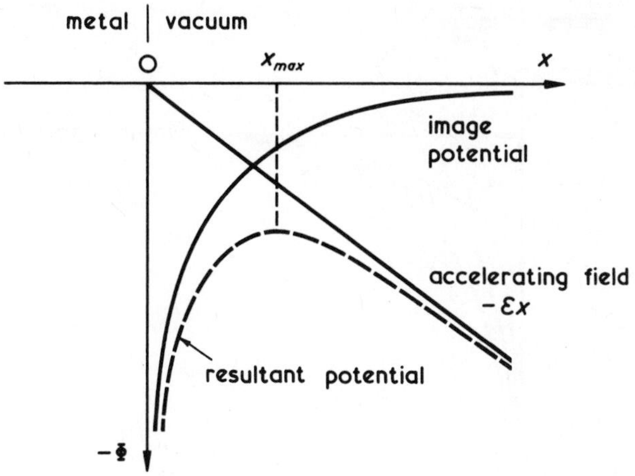

FIG. 5.7 The Schottky effect.

that indicated by the dashed line. Formally we may write the new potential

$$\Phi = \frac{-e^2}{16\pi\varepsilon_0 x} - e\mathcal{E}x \tag{5.105}$$

The maximum in this potential occurs at a value of x given by

$$x_{\max} = \left(\frac{e}{16\pi\varepsilon_0 \mathcal{E}}\right)^{1/2}$$

which represents a *decrease* in the work function $\Delta\phi$ given by

$$\Delta\phi = e\left(\frac{e}{16\pi\varepsilon_0 \mathcal{E}}\right)^{1/2} \tag{5.106}$$

This effect is known as the *Schottky effect*. As an interesting addendum we may note that at very high accelerating potentials, the Schottky barrier becomes so narrow that quantum mechanical tunnelling becomes feasible and we have a phenomenon known as *field emission*. A practical demonstration of this may be found in the *field emission microscope* in which an intense electric field at a very finely etched tip effects field emission and the accelerated electrons impinge on a fluorescent screen (see Gomer[2]).

In order to discuss our simplification (2) we may consider a surface which contains three different crystal faces. We noted already that the form of the image potential is unacceptable at very short distances from the surface, and without enquiring deeply, we may assume an interaction dependent on the type of atomic packing at the surface. This is qualitatively illustrated in in Fig. 5.8. In exaggerated form we see that although the potential of an

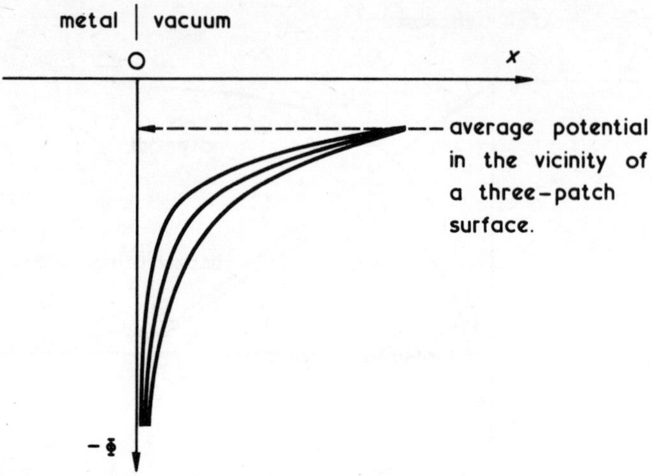

FIG. 5.8 The potential of an electron close to a three-patch surface.

electron escaping from any of the three crystal faces must attain a final constant potential (the work function is a *bulk* property) details of its potential close to a polycrystalline surface is dependent on the crystal face from which it escaped. It is sometimes stated that the work function of a metal depends on the crystal face involved. Since work function is defined as a bulk property this is not very satisfactory and if one is making measurements which provide details of the behaviour of electrons *close* to a surface it is preferable to discuss, perhaps, their *effective work function*. Alternatively we may label this surface sensitive potential, the *surface potential*. Changes in surface potential are frequently used by chemists as a means of studying surface adsorption.

Pauli paramagnetism

As was mentioned in the introduction to this chapter, Pauli's successful explanation[3] of the weak paramagnetism of the alkali metals was of considerable importance in the development of the free electron theory. It is a further example of the predictions of the Fermi–Dirac statistical distribution being different from those expected on classical grounds. In this case, the observed very weak paramagnetism (virtually temperature independent) of the alkali metals is in direct contrast to the temperature dependent strong paramagnetism expected classically. We present a simple description.

In the absence of a magnetic field, we may conveniently represent the one-electron states by two parabolas, one being for electrons of positive spin and one being for electrons of negative spin. In each case, the density of the one-electron states will be $\mathcal{N}(\varepsilon)/2$. This is illustrated in Fig. 5.9(a). When

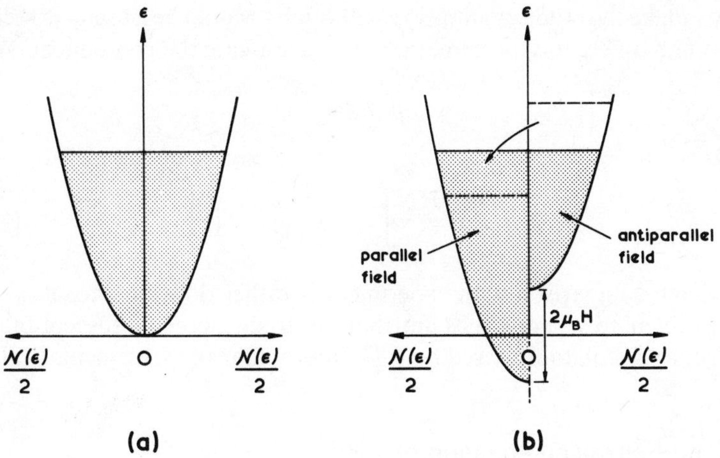

FIG. 5.9 Pauli paramagnetism.

a magnetic field $\boldsymbol{H}$ is switched on, an electron whose energy is ε will be reduced or increased by $\mu_B H$ (μ_B = Bohr magneton) dependent upon whether it is respectively parallel or antiparallel with the magnetic field. Thus the two halves of Fig. 5.9(a) will become displaced by a total of $2\mu_B H$ (see Fig. 5.9(b)). The number of electrons in each distribution will not remain the same, however, since they will equilibrate so that their chemical potentials μ are equal. Thus we may write the following expression for the total magnetic moment $\boldsymbol{\mu}$ of the electron gas

$$\boldsymbol{\mu} = \frac{\mu_B}{2}\left[\int_{-\mu H}^{\infty} f(\varepsilon)\mathcal{N}(\varepsilon + \mu_B H)\,d\varepsilon - \int_{\mu H}^{\infty} f(\varepsilon)\mathcal{N}(\varepsilon - \mu_B H)\,d\varepsilon\right] \quad (5.107)$$

We may expand the density of states functions by use of a Taylor's series.

$$\mathcal{N}(\varepsilon \pm \mu_B H) = \mathcal{N}(\varepsilon) \pm \mu_B H \mathcal{N}'(\varepsilon) \quad (5.108)$$

Using 5.108 in 5.107 and replacing the lower limit of the integrals by zero (since $\mu_B H \ll 1$) we may write

$$\boldsymbol{\mu} = \mu_B{}^2 H \int_0^{\infty} \mathcal{N}'(\varepsilon) f(\varepsilon)\,d\varepsilon \quad (5.109)$$

Equation 5.109 conforms with the standard Fermi–Dirac integral 5.80 which we solve to the quadratic term giving (and this should be checked)

$$\boldsymbol{\mu} = \frac{3\mu_B{}^2 H N \mu^{1/2}}{2\mu_0^{3/2}}\left[1 - \frac{\pi^2}{24}\left(\frac{k_B T}{\mu}\right)^2\right] \quad (5.110)$$

We may make the crude assumption (although it would be a useful exercise to substitute a different approximation and investigate the consequences) that

$$\mu = \mu_0$$

Hence

$$\boldsymbol{\mu} = \frac{3\mu_B{}^2 HN}{2\mu_0}\left[1 - \frac{\pi^2}{24}\left(\frac{k_B T}{\mu_0}\right)^2\right] \tag{5.111}$$

Although comparison with experiment is rather difficult since it is not a direct problem to separate the contributions to the observed susceptibility, it is clear that calculations based on 5.111 are similar to experimental values.

5.5 The Boltzmann equation of state*

To close this chapter we will consider the way in which the free electron theory is a basis for discussion of some of the transport properties of metals. Initially we will derive simply the Boltzmann transport equation or the equation of state.

Let us consider electrons in a metal as constituting a system of particles which is in dynamic equilibrium under some external force e.g. an electric field. Let $f(x_1, x_2, x_3, v_1, v_2, v_3)\, dx_1\, dx_2\, dx_3\, dv_1\, dv_2\, dv_3$ be the number of particles having position co-ordinates in the range x_1 and $x_1 + dx_1$ etc. and velocity coordinates in the range v_1 and $v_1 + dv_1$ etc. in the six-dimensional phase space. The distribution f will not be defined in detail at this stage.

We consider first the time-dependence of f which has two important aspects. First f may vary because particles are moving from one region of space to another and are accelerated in the process by operation of the external force. This continuous variation (the *drift* variation) may be evaluated in the following way. The number of particles that at a time $t + dt$ have drifted to the cell of phase space defined by the coordinates $(x_1, x_2, x_3, v_1, v_2, v_3)$ must be equal to the number that were in the cell defined by the coordinates $[(x_1 - v_1 dt), (x_2 - v_2 dt), (x_3 - v_3 dt), (v_1 - \alpha_1 dt)(v_2 - \alpha_2 dt)(v_3 - \alpha_3 dt)]$ at the time t, where α_1, α_2 and α_3 are the components of the acceleration. It must be noted that the time interval dt is such that collisions are not considered to have had any significant effect on the distribution. Thus the change in the distribution due to drift effects is

$$\begin{aligned}(df)_d = f[&(x_1 - v_1\, dt), (x_2 - v_2\, dt), (x_3 - v_3\, dt), (v_1 - \alpha_1\, dt),\\ &(v_2 - \alpha_2\, dt), (v_3 - \alpha_3\, dt)] - f[x_1, x_2, x_3, v_1, v_2, v_3]\end{aligned} \tag{5.112}$$

* Students may find this section a little involved and it may be skipped at the first reading.

A Taylor expansion of the right-hand side of eqn. 5.112 to the first derivative leads to

$$\left(\frac{\mathrm{d}f}{\mathrm{d}t}\right)_d = -\left(\frac{\partial f}{\partial x_1} v_1 + \frac{\partial f}{\partial x_2} v_2 + \frac{\partial f}{\partial x_3} v_3 + \frac{\partial f}{\partial v_1} \alpha_1 + \frac{\partial f}{\partial v_2} \alpha_2 + \frac{\partial f}{\partial v_3} \alpha_3\right) \tag{5.113}$$

Second, f may vary because of any discontinuous changes in the velocity that accompany collisions. If $\theta(v_1, v_2, v_3, v_1', v_2', v_3')\,\mathrm{d}v_1'\,\mathrm{d}v_2'\,\mathrm{d}v_3'$ is the probability per unit time that a particle will change its velocity from one of components (v_1, v_2, v_3) to one having components in the range v_1' to $v_1' + \mathrm{d}v_1'$ etc. the number of particles a whose velocity changes from (v_1, v_2, v_3) to some others value is

$$a = f(x_1, x_2, x_3, v_1, v_2, v_3) \int \theta(v_1, v_2, v_3, v_1', v_2', v_3')\,\mathrm{d}v_1'\,\mathrm{d}v_2'\,\mathrm{d}v_3' \tag{5.114}$$

Similarly, the number of particles b whose velocity changes *to* (v_1, v_2, v_3) from another value is

$$b = \int f(v_1'', v_2'', v_3'')\theta(v_1'', v_2'', v_3'', v_1, v_2, v_3)\,\mathrm{d}v_1''\,\mathrm{d}v_2''\,\mathrm{d}v_3'' \tag{5.115}$$

Thus the rate of change of f caused by collisions is

$$\left(\frac{\mathrm{d}f}{\mathrm{d}t}\right)_c = b - a \tag{5.116}$$

Hence the total rate of change of f due to drift and collision effects is given by

$$\begin{aligned}\frac{\mathrm{d}f}{\mathrm{d}t} &= \left(\frac{\mathrm{d}f}{\mathrm{d}t}\right)_c + \left(\frac{\mathrm{d}f}{\mathrm{d}t}\right)_d \\ &= 0 \text{ at equilibrium}\end{aligned} \tag{5.117}$$

From eqns 5.113, 5.116 and 5.117 we may write in full

$$\frac{\partial f}{\partial x_1} v_1 + \frac{\partial f}{\partial x_2} v_2 + \frac{\partial f}{\partial x_3} v_3 + \frac{\partial f}{\partial v_1} \alpha_1 + \frac{\partial f}{\partial v_2} \alpha_2 + \frac{\partial f}{\partial v_3} \alpha_3 = b - a$$

or

$$\boldsymbol{v}\,\mathrm{grad}_r\, f + \boldsymbol{\alpha}\,\mathrm{grad}_v\, f = b - a \tag{5.118}$$

Equation 5.118 is called the Boltzmann equation of state or the Boltzmann transport equation (see Seitz)

The term $(\mathrm{d}f/\mathrm{d}t)_c$ may sometimes be expressed in an alternative form by considering a relationship for the rate of relaxation of the system by collision processes if the external force were removed. If f_0 is the thermal equilibrium distribution (this would be $f(\varepsilon)$ for electrons, but we prefer to leave it in a more general form), the time required for the system to relax is given by the following familiar expression

$$(f - f_0)_{t=t} = (f - f_0)_{t=0} \exp\left(-\frac{t}{\tau}\right)$$

where τ is the characteristic *relaxation time* for the process. Alternatively

$$\left(\frac{\mathrm{d}f}{\mathrm{d}t}\right)_c = -\frac{(f - f_0)}{\tau} \tag{5.119}$$

Hence the Boltzmann transport equation may be written

$$\boldsymbol{v}\ \mathrm{grad}_{\boldsymbol{r}}\ f + \boldsymbol{\alpha}\ \mathrm{grad}_{\boldsymbol{v}}\ f = -\frac{f - f_0}{\tau} \tag{5.120}$$

Clearly eqn. 5.120 can be used to describe a number of processes in solids involving a change in the equilibrium distribution of electrons. For example, if we had homogeneous electric fields $\boldsymbol{\mathscr{E}}_1$ and $\boldsymbol{\mathscr{E}}_2$ in X_1 and X_2 directions respectively and a homogeneous magnetic field ($\boldsymbol{H}_3$) in the X_3 direction the accelerations $\boldsymbol{\alpha}_1$ and $\boldsymbol{\alpha}_2$ would be given by an expression of the type

$$m\boldsymbol{\alpha} = -e\boldsymbol{\mathscr{E}} - e\boldsymbol{v} \times \boldsymbol{H}_3 \tag{5.121}$$

and thus substitution in 5.121 leads to (and this should be checked)

$$\frac{\partial f}{\partial x_1} v_1 + \frac{\partial f}{\partial x_2} v_2 + \frac{\partial f}{\partial x_3} v_3 - \frac{\partial f}{\partial v_1}\left(\frac{e\mathscr{E}_1}{m} + \frac{ev_2}{m}\boldsymbol{H}_3\right) - \frac{\partial f}{\partial v_2}\left(\frac{e\mathscr{E}_2}{m} - \frac{ev_1}{m}\boldsymbol{H}_3\right)$$
$$= -\frac{(f - f_0)}{\tau} \tag{5.122}$$

For simplicity we will consider only electrical and thermal conductivity, but the interested reader is urged to consider also the solution of 5.122 (first treated by Lorentz) and electrothermal effects.

Electrical Conductivity

In order to simplify the algebra we will deal with the simplest possible situation: the effect of an electric field $\mathscr{E}_1$ in the X_1 direction. The acceleration produced by $\mathscr{E}_1$ will be given by the equation

$$m\alpha_1 = -e\mathscr{E}_1 \tag{5.123}$$

and the Boltzmann equation reduces to the simple one-dimensional form:

$$\frac{\partial f}{\partial x_1} v_1 - \frac{\partial f}{\partial v_1}\frac{e\mathscr{E}_1}{m} = -\frac{(f - f_0)}{\tau} \tag{5.124}$$

We will make two initial assumptions:

(1) The distribution function f (when the field is present) is related to f_0 (when the field is absent) by the equation

$$f = f_0 + v_1 \chi \tag{5.125}$$

where χ is a small undetermined function that depends on the scalar velocity.

(2) At a constant temperature, f will be independent of position so that

$$\frac{\partial f_0}{\partial x_1} = \frac{\partial f}{\partial x_1} = 0 \tag{5.126}$$

Thus eqn. 5.124 reduces to the form

$$\frac{\partial f}{\partial v_1} \frac{e\mathscr{E}_1}{m} = \frac{v_1 \chi}{\tau} \tag{5.127}$$

By definition, we know that the current J_1 in the direction X_1 is given by

$$J_1 = -ne\langle v_1 \rangle \tag{5.128}$$

where n = the number of electrons per unit volume
and $\langle v_1 \rangle$ = the mean velocity in the direction X_1

We may conveniently express $\langle v_1 \rangle$ in terms of $\boldsymbol{k}$ – space coordinates. Thus

$$\langle v_1 \rangle = \frac{\int_{-\infty}^{\infty}\int_{-\infty}^{\infty}\int_{-\infty}^{\infty} v_1 f \, dk_1 \, dk_2 \, dk_3}{\int_{-\infty}^{\infty}\int_{-\infty}^{\infty}\int_{-\infty}^{\infty} f \, dk_1 \, dk_2 \, dk_3}$$

$$= \frac{V}{4\pi^3 N} \int_{-\infty}^{\infty}\int_{-\infty}^{\infty}\int_{-\infty}^{\infty} v_1 f \, dk_1 \, dk_2 \, dk_3 \tag{5.129}$$

Hence from eqns 5.128 and 5.129

$$J_1 = \frac{-e}{4\pi^3} \int_{-\infty}^{\infty}\int_{-\infty}^{\infty}\int_{-\infty}^{\infty} v_1 f \, dk_1 \, dk_2 \, dk_3 \tag{5.130}$$

Making use of assumption (1) we obtain from 5.125 and 5.130

$$J_1 = \frac{-e}{4\pi^3} \int_{-\infty}^{\infty}\int_{-\infty}^{\infty}\int_{-\infty}^{\infty} v_1^2 \chi \, dk_1 \, dk_2 \, dk_3 \tag{5.131}$$

We may substitute for χ by use of eqn. 5.127. Hence

$$J_1 = \frac{-e^2 \tau \mathscr{E}_1}{4\pi^3 m} \int_{-\infty}^{\infty}\int_{-\infty}^{\infty}\int_{-\infty}^{\infty} v_1 \frac{\partial f}{\partial v_1} \, dk_1 \, dk_2 \, dk_3 \tag{5.132}$$

We now make a further assumption (3) that

$$\frac{\partial f}{\partial v_1} = \frac{\partial f_0}{\partial v_1} \tag{5.133}$$

(which should be acceptable provided $\mathscr{E}_1$ is not too large) and use the identity

$$\frac{\partial f_0}{\partial v_1} = \frac{\partial f_0}{\partial \varepsilon} \cdot \frac{\partial \varepsilon}{\partial v_1} \tag{5.134}$$

where

$$\varepsilon = \frac{m}{2}(v_1{}^2 + v_2{}^2 + v_2{}^2)$$

so that

$$\frac{\partial \varepsilon}{\partial v_1} = mv_1 \tag{5.135}$$

Hence by use of eqns 5.132–5.135, we obtain the equation

$$J_1 = -\frac{e^2\tau\mathscr{E}_1}{4\pi^3}\int_{-\infty}^{\infty}\int_{-\infty}^{\infty}\int_{-\infty}^{\infty} v_1{}^2 \frac{\partial f_0}{\partial \varepsilon}\, \mathrm{d}k_1\, \mathrm{d}k_2\, \mathrm{d}k_3 \tag{5.136}$$

so that provided we can evaluate the volume integral, we should be able to obtain an expression for J_1. We will not attempt a complete solution (although it is not too difficult and those interested should find a way) since it constitutes an example of a so-called general transport integral whose solutions may be expressed as follows (assuming now $f_0 = f(\varepsilon)$).

$$A_n = -\frac{\tau}{4\pi^3}\int_{-\infty}^{\infty}\int_{-\infty}^{\infty}\int_{-\infty}^{\infty} v_1{}^2\varepsilon^{n-1} \frac{\partial f(\varepsilon)}{\partial \varepsilon}\, \mathrm{d}k_1\, \mathrm{d}k_2\, \mathrm{d}k_3 \tag{5.137}$$

$$= -\frac{2\tau}{3mV}\int_0^{\infty} \varepsilon^n \mathscr{N}(\varepsilon) f'(\varepsilon)\, \mathrm{d}\varepsilon \tag{5.138}$$

We note that 5.138 is in form that of a Fermi–Dirac integral (5.81) so may be solved readily to first, second or third order. It is adequate for our purposes to solve 5.136 to first order only, leading simply to

$$J_1 = \frac{e^2\tau\varepsilon_1 N\mu^{3/2}}{\mu_0^{3/2} V} \tag{5.139}$$

Hence, writing $N/V = n$ and making the further simplifying assumption that $\mu = \mu_0$ we obtain the very simple expression

$$J_1 = \frac{e^2\tau\mathscr{E}_1 n}{m} \tag{5.140}$$

which is in the form of Ohm's Law. The electrical conductivity σ will thus be given by

$$\sigma = \frac{e^2 \tau n}{m} \tag{5.141}$$

The form of 5.141 makes sense! We expect the charge transported to be proportional to en; the factor e/m enters because the acceleration in an electric field is proportional to e and inversely proportional to m. The relaxation time τ describes the 'free time' during which the field can act on the electron. We might qualitatively use conductance measurements in order to calculate values of τ; it would be instructive to make such a calculation for a number of metals. Further it might be instructive to reconsider the use of the approximation $\mu = \mu_0$ by an alternative application of 5.89.

Thermal conductivity

In principle it is not difficult to use a formally similar approach in order to deal with the problem of thermal conductivity. We assume that the thermal conductivity of a metal is due only to its electrons without any lattice contributions. Thus in the one-dimensional case, a thermal gradient dT/dx_1 leads to a thermal current density. Since the gradient produces a finite drift velocity a small electric field must be set up to counterbalance this effect at equilibrium.

In our discussion of isothermal conductivity we could assume that

$$\frac{\partial f}{\partial x_1} = \frac{\partial f_0}{\partial x_1} = 0.$$

The effect of the necessary introduction of the finite term $\partial f_0/\partial x_1$ in a treatment of thermal conductivity may be seen most readily by first expanding this term assuming the Fermi–Dirac distribution function. Thus differentiation of f_0 leads to the equation

$$\frac{\partial f_0}{\partial x_1} = \frac{\partial f_0}{\partial \varepsilon}\left[\frac{\mu}{T} - \frac{\varepsilon}{T} - \frac{\partial \mu}{\partial T}\right]\frac{\partial T}{\partial x_1} \tag{5.142}$$

which may be expressed conveniently in the form

$$\frac{\partial f_0}{\partial x_1} = -\frac{\partial f_0}{\partial \varepsilon}\left(\frac{\varepsilon}{T}\right)\frac{\partial T}{\partial x_1} - \frac{\partial f_0}{\partial \varepsilon} T \frac{\partial}{\partial T}\left(\frac{\mu}{T}\right)\frac{\partial T}{\partial x_1} \tag{5.143}$$

Substitution from eqns 5.123, 5.125, 5.133–5.135 and 5.143 in the transport equation, which is of the form

$$v_1 \frac{\partial f_0}{\partial x_1} + \alpha_1 \frac{\partial f_0}{\partial v_1} = -\frac{(f - f_0)}{\tau} \tag{5.144}$$

leads to

$$-\frac{e\mathscr{E}_1}{m}\frac{\partial f_0}{\partial \varepsilon} - \frac{\partial T}{\partial x_1}\cdot\frac{\partial f_0}{\partial \varepsilon}\left[-\frac{\varepsilon}{T} - \frac{T\partial}{\partial T}\left(\frac{\mu}{T}\right)\right] = \frac{\chi}{\tau} \tag{5.145}$$

Substitution of 5.145 in 5.131 leads to an equation similar to 5.136 expect, of course, the extra terms of 5.145 are involved:

$$J_1 = -\frac{e\tau}{4\pi^3}\int_{-\infty}^{\infty}\int_{-\infty}^{\infty}\int_{-\infty}^{\infty}$$
$$\times\left\{v_1{}^2 e\mathscr{E}_1\frac{\partial f_0}{\partial \varepsilon} - v_1{}^2\frac{\partial T}{\partial x_1}\left[-\frac{\varepsilon}{T} - T\frac{\partial}{\partial T}\left(\frac{\mu}{T}\right)\right]\frac{\partial f_0}{\partial \varepsilon}\right\} \mathrm{d}k_1\,\mathrm{d}k_2\,\mathrm{d}k_3 \tag{5.146}$$

Now we use the condition that $J_1 = 0$ in order to calculate the field $\mathscr{E}_1$.

This condition leads to

$$A_1 e\mathscr{E}_1 + A_1 T\frac{\partial}{\partial x_1}\left(\frac{\mu}{T}\right) + A_2\frac{1}{T}\frac{\partial T}{\partial x_1} = 0 \tag{5.147}$$

where A_1 and A_2 are the transport integrals (5.137) of respective orders 1 and 2. Hence

$$e\mathscr{E}_1 = -T\frac{\partial}{\partial x_1}\left(\frac{\mu}{T}\right) - \frac{A_2}{A_1}\frac{1}{T}\frac{\partial T}{\partial x_1} \tag{5.148}$$

If Q_1 is the thermal current density which we may define by

$$Q_1 = n\varepsilon\langle v_1\rangle$$

it follows by comparison with 5.128 that

$$Q_1 = -\frac{J_1\varepsilon}{e}$$

Hence from 5.146 we may write the following expression for Q_1

$$Q_1 = -A_2 e\mathscr{E}_1 - A_2 T\frac{\partial}{\partial x_1}\left(\frac{\mu}{T}\right) - A_3\frac{1}{T}\frac{\partial T}{\partial x_1} \tag{5.149}$$

which on substituting the value of $\mathscr{E}_1$ appropriate to the condition that $J_1 = 0$ (5.148) leads to an expression for the thermal conductivity $\mathscr{K}_e$

$$\mathscr{K}_e = \frac{Q_1}{-\partial T/\partial x_1} = \frac{A_1 A_3 - A_2{}^2}{TA_1} \tag{5.150}$$

Evaluation of A_1, A_2 and A_3 to the first order of approximation will quickly demonstrate that the temperature independent terms cancel and the parts of the expression containing second order terms must therefore be included leading to

$$\mathcal{K}_e = \frac{N\pi^2 k_B{}^2 \tau T}{3mV}$$

$$= \frac{n\pi^2 k_B{}^2 \tau T}{3m} \qquad (5.151)$$

In pure metals the electronic contribution is dominant at all temperatures, the thermal conductivity being typically one or two orders of magnitude higher than that of non-conducting solids. The *Wiedemann–Franz* law states that for metals the ratio of the thermal conductivity to the electrical conductivity is directly proportional to the temperature, the value of the constant of proportionality (the Lorenz number L) being independent of the metal. According to eqns 5.141 and 5.151

$$L = \frac{\mathcal{K}_e}{\sigma T} = \frac{\pi^2}{3}\left(\frac{k_B}{e}\right)^2$$

$$= 2.45 \mathrm{J^2 A^{-2} S^{-2} K^{-2}} \qquad (5.152)$$

Experimental results (see Kittel) are in remarkably good agreement with this result.

It should be noted that other transport properties of metals may be treated by techniques analogous to those outlined above and the interested student is referred to Seitz and to Wilson.[4] A reaction, not without substance, to the above treatment of electrical and thermal conductivity might be that qualitative plausibility arguments (see Kittel) can lead to expressions for σ and $\mathcal{K}_e$ without recourse to prolonged analysis. However, the decision to include a more complete treatment was prompted by its more general utility, which hopefully the reader will explore. Finally it will have been observed that we have not mentioned phonon–electron scattering (as outlined in Chapter 4) which is the major contribution to the electrical resistivity at room temperature. The reader will find a further discussion of this in Kittel.

References

(1) A. Sommerfeld, *Z. Physik* **47**, 1 (1928)
(2) R. Gomer, *Field Emission and Field Ionisation* (Harvard University Press, 1961)
(3) W. Pauli, *Z. Physik* **41**, 81 (1927)
(4) A. H. Wilson, *The Theory of Metals*, 2nd edition (Cambridge University Press, 1954)

General

A. H. Wilson (*see* (4))

6

Electrons in a Periodic Potential

6.1 A qualitative discussion of conductors, semiconductors and insulators

In Chapter 5 our model for a metal was a very simple one, according to which electrons (subject to certain boundary conditions) move in a constant potential. Some success was achieved, but one major question (at least) remains unanswered: why are some materials good electrical conductors while some materials are good electrical insulators? There is no reason, in principle, why we should not apply our free electron theory to an element like carbon (in the diamond form) and expect high electrical and thermal conductivities.

In order to overcome such anomalies and in order to begin a discussion of a fascinating range of materials known as semiconductors, we have to extend our free electron model by incorporating the variation in potential due to the positive ion cores; this extension results in so-called *energy gaps* or *band gaps* which are regions of energy which do not correspond to allowed solutions of the Schrödinger equation. In the simple one-dimensional case, we may appreciate the effect of including a *periodic potential* (the periodicity being that of the lattice a) by inspection of Fig. 6.1. We see that in contrast to free electrons, band gaps occur at $k = \pm n\pi/a$. In particular, we note that electrons cannot have energies in the range ε_1 to ε_2. Although we have anticipated later results without justification in order to present Fig. 6.1, we may demonstrate nevertheless the physical plausibility of the band gaps at $k = \pm n\pi/a$, by considering Bragg reflexion in a one-dimensional lattice. We may represent the one-dimensional reciprocal lattice numbers by

$$G = \frac{2n\pi}{a} \tag{6.1}$$

so that the Bragg condition (3.61) in one dimension becomes

$$k = \pm G/2$$

$$= \pm \frac{n\pi}{a} \tag{6.2}$$

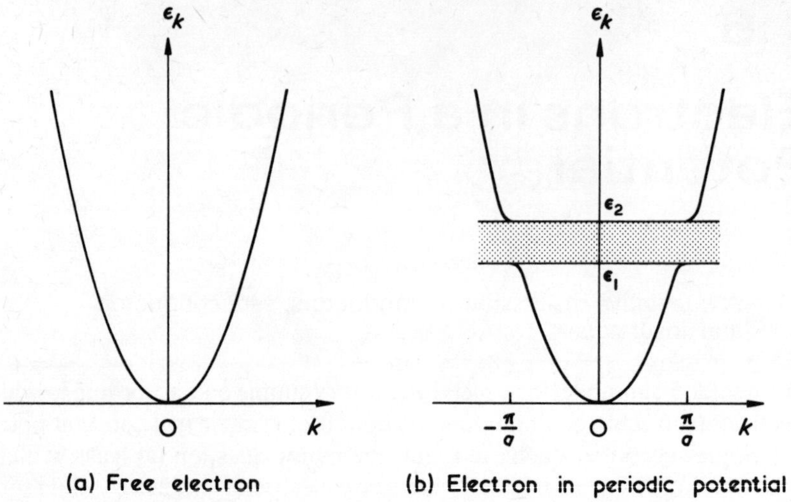

FIG. 6.1 The effect of a one-dimensional periodic potential.

Examining the simple case when $n = 1$, we see that eqn. 6.2 describes the boundary of the first Brillouin zone and the condition for first order Bragg reflexion. We have noted this in previous discussion and what we have tended separately to describe as reciprocal lattice space and $\boldsymbol{k}$-space are, of course, identical. Thus an electron whose wave number k satisfies the condition 6.2 is Bragg reflected and cannot propagate through the lattice. In fact an electron of $k = \pm\pi/a$, instead of being described by a travelling wave is described by standing waves which are linear combinations of the former, i.e.

$$\psi_{\text{standing}} = C\left[\exp\left(\frac{\pi i x}{a}\right) \pm \exp\left(-\frac{\pi i x}{a}\right)\right] \tag{6.3}$$

or

$$\psi^{+} = 2C\cos\frac{\pi x}{a} \tag{6.4}$$

and

$$\psi^{-} = 2Ci\sin\frac{\pi x}{a} \tag{6.5}$$

Following Kittel, we may conveniently represent the probability density ($\psi^*\psi$) of the electrons ψ^+ and ψ^- together with that of the travelling wave in Fig. 6.2. The probability density for the travelling wave is a *constant*, that of the standing wave ψ^- is a maximum *between* the ion cores and that of the standing wave ψ^+ is a maximum *at* the ion cores. Since the potential energy of

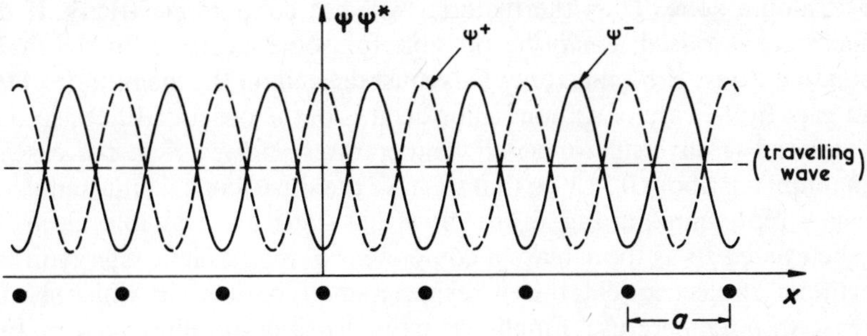

FIG. 6.2 Probability density of an electron in a one-dimensional lattice.

an electron will tend to $-\infty$ at the ion cores, we expect the potential energy of ψ^+ to be the lowest. A similar argument suggests that the potential energy of ψ^- will be the highest and that of the travelling wave somewhere in between. The difference between ψ^+ and ψ^- will then correspond to the energy gap $\varepsilon_2 - \varepsilon_1$.

We now consider whether or not the existence of band gaps will help in our distinguishing materials which have (1) a high electrical conductivity decreasing with temperature (typical metals), (2) a low electrical conductivity increasing with temperature (semiconductors Si, Ge etc.) and (3) no significant electrical conductivity (typical insulators). Assuming that our one-dimensional model is appropriate for three-dimensional systems, we may examine several possible situations. We envisage first the situation where the electrons may *just* be accommodated up to a wave number of $k = \pm\pi/a$ (we might say the Brillouin zone is filled) at 0 K. We indicate this by the shading in Fig. 6.3(a). Application of an electric field cannot increase the kinetic energy of the electrons since there are no empty states available within the

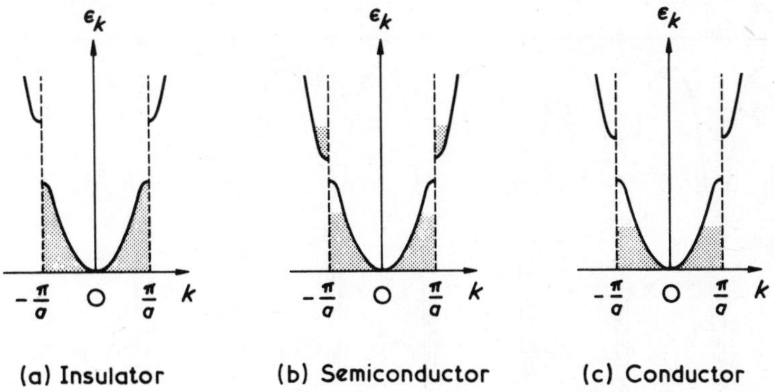

FIG. 6.3 Insulators, semiconductors and conductors.

first Brillouin zone. Thus the material will not conduct electricity. If the temperature is raised it *may* be possible for some electrons to thermally populate the next Brillouin zone. This must depend on the magnitude of the band gap. In the case of sodium chloride, it is of the order of 10 eV, so that the material is an insulator at all temperatures. However, in the case of germanium it is about 0·7 eV so that we have the situation at a finite temperature as is represented in Fig. 6.3(b). Now that states are available, electrons may be excited so as to increase a component of their kinetic energy in the direction of an electric field; thus we expect a small conductivity which should increase with temperature. Finally we have the situation illustrated by Fig. 6.3(c) in which we have (typically) a half-filled band. Here the high conductivity of a metal is expected.

Although the general arguments presented in the previous paragraph are not incorrect and hopefully begin to provide some flavour of the problems involved, we clearly require a more convincing formal basis for discussion. Several sections will be devoted to this end.

6.2 The one-dimensional periodic potential

Bloch functions, Brillouin zones and band gaps

We represent the problem in Fig. 6.4, knowing from previous discussion that if $V(x) = 0$, the free electron wave function may be written

$$\psi_k = \left(\frac{1}{L}\right)^{1/2} \exp(ikx)$$

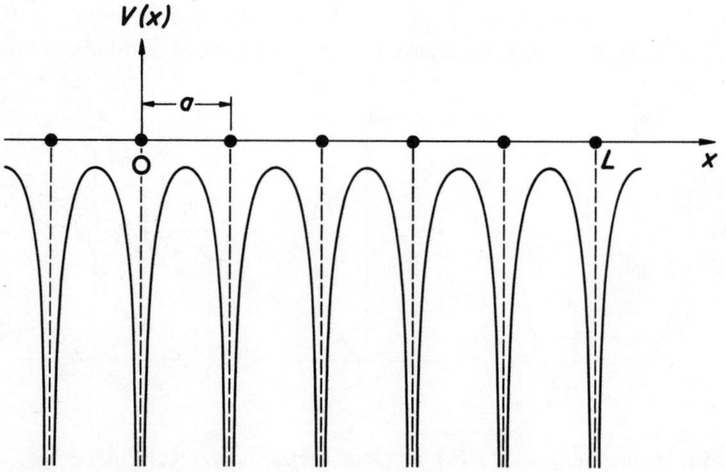

FIG. 6.4 The one-dimensional periodic potential.

where

$$k = \frac{2\pi n}{L} \qquad (n = 0, \pm 1, \pm 2 \text{ etc.})$$

We wish now to solve the Schrödinger equation which includes the period potential $V(x)$ which may be defined by means of the lattice constant a

$$V(x) = V(x + a) \tag{6.6}$$

The appropriate equation may be written

$$-\frac{\hbar^2}{2m}\frac{\mathrm{d}^2\psi}{\mathrm{d}x^2} + V(x)\psi = \varepsilon\psi \tag{6.7}$$

and the periodic boundary conditions are

$$\psi(x) = \psi(x + Na) \tag{6.8}$$

where obviously $N = L/a$.

Equation 6.7 is most easily solved by reference to the properties of *simultaneous eigenfunctions*, so for the unfamiliar reader, we deviate briefly from our main line of development to discuss this topic. We have already alluded to simultaneous eigenfunctions when we showed (eqns 5.44 and 5.45) that the eigenfunctions of the free electron Hamiltonian are also eigenfunctions of the momentum operator leading to a momentum $p = \hbar k$. We thus say that p is also a constant of motion. Operators which represent constants of motion may also be defined by a commutator relationship, readily developed as follows. If the operator α represents a constant of motion, the equations

$$H\psi = E\psi \tag{6.9}$$

and

$$\alpha\psi = \beta\psi \tag{6.10}$$

(where β is the eigenvalue)

must be satisfied by the same ψ. Operating on each side of 6.9 by the operator α, and on each side of 6.10 by the operator H

$$\alpha H\psi = \alpha E\psi = E\alpha\psi = E\beta\psi \tag{6.11}$$

and

$$H\alpha\psi = H\beta\psi = \beta H\psi = \beta E\psi \tag{6.12}$$

From 6.11 and 6.12

$$\alpha H\psi = H\alpha\psi$$

or

$$(\alpha H - H\alpha)\psi = 0 \tag{6.13}$$

In other words, the operators α and H commute; this is expressed normally by the following formalism

$$[\alpha, H] = 0 \tag{6.14}$$

and we may say that *any operator which represents a constant of motion commutes with the Hamiltonian.*

After this diversion, let us return to the solution of eqn. 6.7. Let Γ be an operator which effects the change

$$x \rightarrow x + a$$

so that operating on the left hand side of 6.7 leads to

$$\Gamma H\psi(x) = \left[-\frac{\hbar^2}{2m}\frac{\mathrm{d}^2}{\mathrm{d}(x+a)^2} + V(x+a) \right]\psi(x+a) \tag{6.15}$$

But owing to the condition 6.6 and the fact that

$$\frac{\mathrm{d}}{\mathrm{d}(x+a)} = \frac{\mathrm{d}}{\mathrm{d}x}$$

we have

$$\begin{aligned} \Gamma H\psi(x) &= H\psi(x+a) \\ &= H\Gamma\psi(x) \end{aligned} \tag{6.16}$$

Thus we may write the commutator relationship

$$[\Gamma, H] = 0 \tag{6.17}$$

so that the two operators have simultaneous eigenfunctions. Hence

$$\begin{aligned} \Gamma\psi(x) &= \gamma\psi(x) \\ &= \psi(x+a) \end{aligned} \tag{6.18}$$

or

$$\psi(x+a) = \gamma\psi(x) \tag{6.19}$$

where γ is a constant (of motion but without classical analogue). Generally we may write

$$\psi(x+na) = \gamma^n\psi(x) \tag{6.20}$$

From eqns 6.8 and 6.20 it follows that

$$\psi(x+Na) = \gamma^N\psi(x) = \psi(x) \tag{6.21}$$

Hence from 6.21

$$\gamma^N = 1 \tag{6.22}$$

so that γ must be one of the roots

$$\gamma = \exp\left(\frac{2\pi in}{N}\right) \qquad (n = 0, \pm 1, \pm 2 \text{ etc.}) \tag{6.23}$$

If we define

$$k = \frac{2\pi n}{L} \quad \text{(as in eqn. 5.49)}$$

$$= \frac{2\pi n}{Na} \qquad (n = 0, \pm 1, \pm 2 \text{ etc.}) \tag{6.24}$$

it follows from eqn. 6.19 that

$$\psi(x + a) = \exp(ika)\psi(x) \tag{6.25}$$

frequently referred to as the *Bloch condition.*

If we choose a function u_k which has the periodicity of the lattice i.e.

$$u_k(x) = u_k(x + a) \tag{6.26}$$

we note that eqn. 6.25 is consistent with a solution of the type

$$\psi(x) = u_k(x)\exp(ikx) \tag{6.27}$$

since from eqn. 6.27

$$\begin{aligned} \psi(x + a) &= u_k(x + a)\exp[ik(x + a)] \\ &= \exp(ika)u_k(x)\exp(ikx) \\ &= \exp(ika)\psi(x) \end{aligned} \tag{6.28}$$

Equation 6.27 is known as a *Bloch function* or the statement that the wave functions describing an electron in a periodic potential are equivalent to plane travelling waves modulated by a function which has the periodicity of the lattice is referred to as the *Bloch theorem*[1].

Even at this early stage of discussion we may make some noteworthy comments about Bloch functions. It will be observed immediately on inspection of 6.27, that k is not uniquely defined by this equation since we may write

$$\exp(ikx)u_k(x) = \exp\left[i\left(k + \frac{2\pi l}{a}\right)x\right]\exp\left(-\frac{2\pi ilx}{a}\right)u_k(x)$$

(l = any integer) since $\exp(-2\pi ilx/a)$, which is of course the same as $\exp(ilGx)$, has the necessary lattice periodicity. It is therefore possible (as we did in Chapter 4) to restrict the values of k to an interval of length $2\pi/a$ without loss of information. For convenience, we normally take the first Brillouin zone ($-\pi/a \leq k \leq +\pi/a$) as the fundamental domain. We may rewrite 6.27 in a slightly modified form to emphasise that for any chosen value of k in the first Brillouin zone there is an infinite number of states differing by values of l.

$$\psi_{kl}(x) = \exp(ikx)u_{kl}(x) \tag{6.29}$$

$$\text{where } u_{kl}(x) = u_{kl}(x + a) \tag{6.30}$$

As a simple, if somewhat artificial example, we may represent free electron wave functions in this way. By letting $V(x) = 0$, we may obtain the following Bloch functions in the range $-\pi/a \leq 0 \leq +\pi/a$ and corresponding eigenvalues. For $0 < k \leq +\pi/a$

$$\psi_{kl}(x) = \exp(ikx)\exp\left(-\frac{2\pi ilx}{a}\right) \tag{6.31}$$

$$\varepsilon_k = \frac{\hbar^2}{2m}\left(k - \frac{2\pi l}{a}\right)^2 \tag{6.32}$$

and for $-\pi/a \leq k < 0$

$$\psi_{kl}(x) = \exp(ikx)\exp\left(\frac{2\pi ilx}{a}\right) \tag{6.33}$$

$$\varepsilon_k = \frac{\hbar^2}{2m}\left(k + \frac{2\pi l}{a}\right)^2 \tag{6.34}$$

The ε_k *versus* k curves for several values of l are shown in Fig. 6.5(a). Each value of l defines a Brillouin zone in the *reduced zone scheme representation.* We will return to this representation shortly.

Having noted that Bloch waves are no more than free electron waves modulated by the periodic function $u_k(x)$, we may ask the question: what effect will $u_k(x)$ have on a 'free' wave packet built up from stationary state wave functions. A wave packet constructed from wave functions (6.27) with k values in a *small* interval δk will travel through the lattice at a constant velocity

$$v_g = \frac{\mathrm{d}\omega}{\mathrm{d}k} = \frac{1}{\hbar}\frac{\mathrm{d}\varepsilon}{\mathrm{d}k}$$

if one assumes that the modulation function is independent of k in the range δk. Recalling that the extent of the wave packet $\delta x \simeq 1/\delta k$ (those not familiar with this may find a clear discussion, for example, in Scharff[2]) we might expect

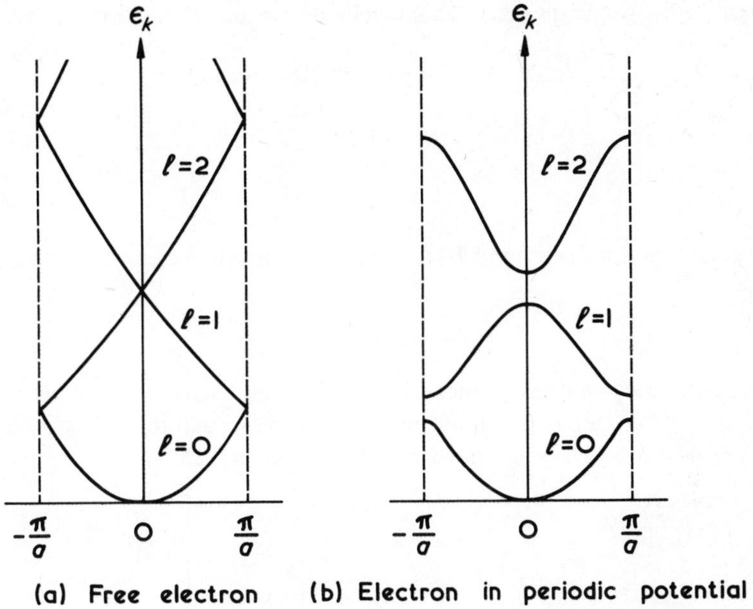

FIG. 6.5 Electron energies in the reduced zone scheme representation.

it to be long compared with the lattice periodicity a, so that the lattice modulation should have little influence. Qualitatively, therefore, provided $d\varepsilon/dk \neq 0$, we would expect a wave packet to propagate freely through a periodic lattice; any *deviations* from periodicity produce scattering.

In order to demonstrate the existence of band gaps, we take advantage of the assumption implicit in Fig. 6.4, that the potential is symmetric around $x = 0$ (we choose the fundamental domain $-a/2 \leq x \leq a/2$)

$$\text{i.e.} \quad V(x) = V(-x) \tag{6.35}$$

If we define a *parity operator* P by the equation

$$P\psi(x) = \psi(-x) \tag{6.36}$$

it may be immediately demonstrated that for an electron in the potential defined by 6.35

$$[P, H] = 0 \tag{6.37}$$

Therefore P is a constant of motion and it is thus possible to find a set of simultaneous eigenfunctions of P and H.

$$P\psi(x) = y\psi(x) \tag{6.38}$$

It is easy to evaluate the eigenvalue y since P^2 is equal to the identity operator

$$\text{i.e.}\quad P^2\psi(x) = \psi(x) \\ = y^2\psi(x) \tag{6.39}$$

or

$$y = \pm 1 \tag{6.40}$$

The eigenfunctions of P (and H) are thus even or odd functions and we may write

$$\psi(x) = A\psi_1(x) + B\psi_2(x) \tag{6.41}$$

where ψ_1 and ψ_2 may be assumed to be respectively even and odd functions of x. The wave function of 6.41 must be a Bloch wave so that the following continuity conditions at $x = a/2$ must be satisfied (from 6.25)

$$A\psi_1\left(\frac{a}{2}\right) + B\psi_2\left(\frac{a}{2}\right) = \exp(ika)\left[A\psi_1\left(-\frac{a}{2}\right) + B\psi_2\left(-\frac{a}{2}\right)\right] \tag{6.42}$$

and

$$A\psi'_1\left(\frac{a}{2}\right) + B\psi'_2\left(\frac{a}{2}\right) = \exp(ika)\left[A\psi'_1\left(-\frac{a}{2}\right) + B\psi'_2\left(-\frac{a}{2}\right)\right] \tag{6.43}$$

Making use of the parity of $\psi_1(x)$ and $\psi_2(x)$ in 6.42 and 6.43 we obtain

$$A\psi_1\left(\frac{a}{2}\right)[1 - \exp(ika)] + B\psi_2\left(\frac{a}{2}\right)[1 + \exp(ika)] = 0 \tag{6.44}$$

and

$$A\psi'_1\left(\frac{a}{2}\right)[1 + \exp(ika)] + B\psi'_2\left(\frac{a}{2}\right)[1 - \exp(ika)] = 0 \tag{6.45}$$

Non-trivial solutions for A and B in 6.44 and 6.45 depend on the condition that

$$\begin{vmatrix} \psi_1\left(\frac{a}{2}\right)[1 - \exp(ika)] & \psi_2\left(\frac{a}{2}\right)[1 + \exp(ika)] \\ \psi'_1\left(\frac{a}{2}\right)[1 + \exp(ika)] & \psi'_2\left(\frac{a}{2}\right)[1 - \exp(ika)] \end{vmatrix} = 0 \tag{6.46}$$

or

$$\cos ka = \frac{\dfrac{\psi_1'\left(\frac{a}{2}\right)}{\psi_1\left(\frac{a}{2}\right)} + \dfrac{\psi_2'\left(\frac{a}{2}\right)}{\psi_2\left(\frac{a}{2}\right)}}{\dfrac{\psi_1'\left(\frac{a}{2}\right)}{\psi_1\left(\frac{a}{2}\right)} - \dfrac{\psi_2'\left(\frac{a}{2}\right)}{\psi_2\left(\frac{a}{2}\right)}} \tag{6.47}$$

Since physically acceptable solutions of the original wave equation must contain values of k which are wholly real, the function $\cos ka$ is limited to the range between -1 and $+1$; energy values corresponding to values of k outside those limits are *inaccessible* and correspond to band gaps. The function $\cos ka$ is periodic in k with a period of $2\pi/a$ so that as in our previous discussion we may limit the k values to the interval $-\pi/a \leq k \leq \pi/a$ according to the reduced zone scheme representation [see Fig. 6.5(b)].

Before leaving our qualitative discussion of band gaps in the linear lattice we may finally calculate very simply the number of electron states in a zone. According to eqn. 6.24, which imposes periodic boundary conditions on the Bloch waves, the allowed wave numbers in the first Brillouin zone are

$$k = 0, \pm \frac{2\pi}{L} \pm \frac{4\pi}{L} \cdots \frac{N\pi}{L} \tag{6.48}$$

We terminate the series at $k = N\pi/L(=\pi/a)$ since it constitutes a zone boundary noting also that $-N\pi/L$ is not to be taken as *independent* since it is related to $k = N\pi/L$ by the reciprocal lattice number $2\pi/a$. We see, therefore, that the total number of allowed k values in 6.48 is N. In other words one allowed level may be formally associated with each primitive cell of the linear lattice, so taking the two possible spin orientations into account, the first Brillouin zone can accommodate a total of two electrons per lattice point (this result is also true in three dimensions). Thus we might argue on the basis of our one-dimensional model, that, as expected, the alkali metals (supplying one electron per atom and thus half-filling the zone) should conduct. By the same token, we might expect magnesium to be an insulator! Clearly the one-dimensional model has limitations.

The Kronig–Penney model

Having established in outline the essential features of the behaviour of an electron in an undefined periodic potential, it is a pedagogically valuable exercise to consider a simple one-dimensional potential which allows an

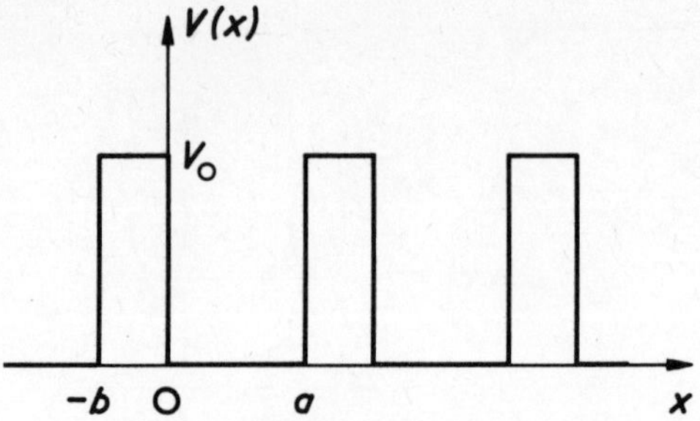

FIG. 6.6 The Kronig–Penney potential.

exact calculation of the ε-k relationship. Our treatment will be rather superficial but it is recommended that students should read the original paper by Kronig and Penney.[3] The assumed potential contains periodic square wells as shown in Fig. 6.6. Thus we see that the period of the potential is $a + b$. In the region $0 < x < a$ the potential energy is assumed to be zero and in the region $-b < x < 0$, the potential energy is assumed to be V_0. The appropriate Schrödinger equations for the two regions are

$$\frac{\hbar^2}{2m}\frac{d^2\psi}{dx^2} + \varepsilon\psi = 0 \qquad \text{for} \quad 0 < x < a \tag{6.49}$$

and

$$\frac{\hbar^2}{2m}\frac{d^2\psi}{dx^2} + (\varepsilon - V_0)\psi = 0 \qquad \text{for} \quad -b < x < 0 \tag{6.50}$$

Following the original formalism, we define two real quantities β and γ by the equations

$$\beta = \frac{2m\varepsilon}{\hbar^2} \tag{6.51}$$

and

$$\gamma = \frac{2m(V_0 - \varepsilon)}{\hbar^2} \tag{6.52}$$

and make use of the assumption that solutions of the wave eqns 6.49 and 6.50 must be Bloch functions of the type

$$\psi(x) = u(x)\exp(\pm ikx) \tag{6.53}$$

Hence substitution in 6.49 and 6.50 leads to the equations

$$\frac{d^2u_1}{dx^2} + 2ik\frac{du_1}{dx} - u_1(k^2 - \beta^2) = 0 \tag{6.54}$$

for

$$0 < x < a$$

and

$$\frac{d^2u_2}{dx^2} + 2ik\frac{du_2}{dx} - u_2(k^2 - \gamma^2) = 0 \qquad (6.55)$$

for

$$-b < x < 0$$

Solutions of 6.54 and 6.55 may be written in the form

$$u_1 = A\exp[i(\beta - k)x] + B\exp[-i(\beta + k)x] \qquad (6.56)$$

and

$$u_2 = C\exp[i(\gamma - k)x] + D\exp[-i(\gamma + k)x] \qquad (6.57)$$

where A, B, C and D are arbitrary constants

The requirement of continuity for the wave function and its derivative demands that the functions u_1 and u_2 satisfy these very conditions since $\exp(ikx)$ is a well-behaved function. Thus

$$u_1(0) = u_2(0);\ u_1(a) = u_2(-b)$$

$$\left(\frac{du_1}{dx}\right)_{x=0} = \left(\frac{du_2}{dx}\right)_{x=0};\ \left(\frac{du_1}{dx}\right)_{x=a} = \left(\frac{du_2}{dx}\right)_{x=-b} \qquad (6.58)$$

The continuity conditions (6.58) lead to four homogeneous equations in A, B, C and D so that the determinant of their coefficients may be equated to zero. Expansion of the determinant leads to the equation

$$\frac{\gamma^2 - \beta^2}{2\beta\gamma}\sinh\gamma b\sin\beta a + \cosh\gamma b\cos\beta a = \cos k(a + b) \qquad (6.59)$$

Equation 6.59 may be simplified in the limit of $b \to 0$ and $V_0 \to \infty$; here only the first terms in the series expansions of the sinh and cosh terms need to be retained. If the limit is chosen so that the product $\gamma^2 b$ remains finite, 6.59 becomes

$$\frac{P\sin\beta a}{\beta a} + \cos\beta a = \cos ka \qquad (6.60)$$

where

$$P = \lim_{\substack{b\to 0\\ \gamma\to\infty}}\left[\frac{\gamma^2 ab}{2}\right] \qquad (6.61)$$

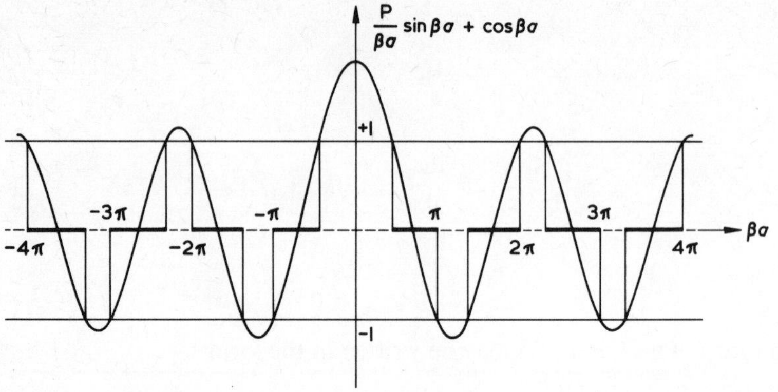

FIG. 6.7 Forbidden energy values according to the Kronig–Penney model.

To discuss the roots of the transcendental eqn. 6.60 it is convenient to plot its left hand side as a function of βa; following Kronig and Penney we use an arbitrary value of $P = 3\pi/2$ (see Fig. 6.7). The values of βa satisfying 6.61 are obtained as projections on to the βa axis of the intersections of the curve with a straight line drawn at the distance $\cos ka$ parallel to the axis. Since $\cos ka$ must lie between -1 and $+1$, and since the ordinates of the maxima have magnitudes greater than unity then allowed energy values of an electron are separated by forbidden intervals.

It is of interest to look at the influence of P on the energy spectrum. If P is made equal to zero, the curve of Fig. 6.7 becomes that of $\cos \beta a$, the forbidden intervals disappear and we have the continuous energy spectrum of a free electron. If instead, P is increased, the ratio of the lengths of the forbidden intervals to that of the allowed regions decreases. As P approaches infinity, the allowed regions on the βa axis tend to points at $n\pi$ ($n = \pm 1$, ± 2 etc.); the energy spectrum has become discrete. Returning to eqn. 6.60 and Fig. 6.7 we observe that the energy discontinuities occur at $k = n\pi/a$ ($n = \pm 1, \pm 2 \cdots$) which values define the boundaries between the Brillouin zones. We illustrate this in Fig. 6.8 in a way different from that used earlier (Fig. 6.5): instead of restricting k to a particular range we allow an unlimited range of values. The representation is referred to as the *extended zone scheme* and is equivalent to that of Fig. 6.5(b) since we know that Bloch functions have a periodicity of $2\pi l/a$. Thus translation by the reciprocal lattice number $2\pi l/a$ leads from Fig. 6.8 to Fig. 6.5(b).

One further aspect of Fig. 6.7 should be considered: it is not self evident that the forbidden energy regions must necessarily continue for all values of βa. We discuss briefly this question using the approach of Jones[4]. Let

$$\beta a = n\pi + \delta n \tag{6.62}$$

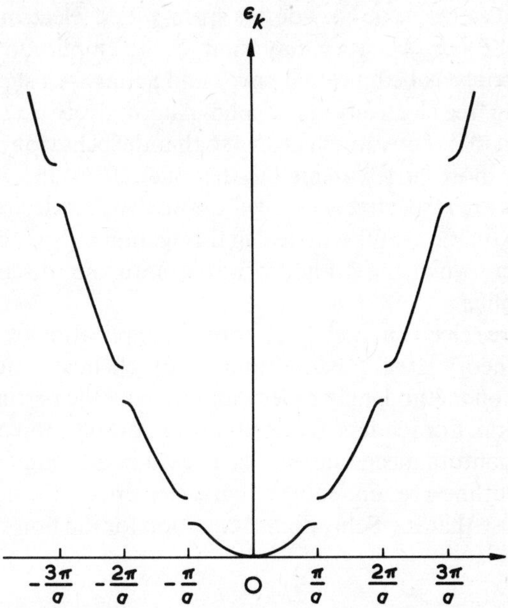

FIG. 6.8 Energy of an electron in the Kronig–Penney lattice (extended zone scheme).

where δn is a very small positive number. A little algebraic manipulation (which should be done) will show that substitution for βa as defined in 6.62 in the left-hand side of eqn. 6.60 may be written

$$(-1)^n\left(1 + \frac{P\delta n}{n\pi}\right) \tag{6.63}$$

which is >1 when n is an even integer
<-1 when n is an odd integer

Thus the band structure exists for all energies however large.

It must have become obvious that in this and the previous section, we have been somewhat repetitive. We have considered the Kronig–Penney potential using the traditional treatment, whereas it should have been possible by an appropriate symmetrical definition of the potential to proceed from eqn. 6.47. The latter approach was purposely avoided in order to present the historically important analysis of Kronig and Penney and to give the reader the opportunity to follow independently the alternative route.

*The nearly-free electron model**

After a brief consideration of the specific periodic potential of Kronig and Penney we now return to a more general approach. One may appreciate

* The subsequent simple treatment is very similar to that presented by Raimes.[5]

that there are two extreme viewpoints from which electrons in a periodic potential may be seen. First we may think of the amplitude of the potential of the atomic cores as being rather small and acting as a slightly perturbing influence on the free electrons in the solid. Alternatively we may think of the solid as a fusion of its constituent atoms so that the behaviour of the electrons in the former is more closely related to that of electrons in a free atom. These two viewpoints are respectively labelled the nearly-free electron and the tight binding approximations. We will deal in this section only with the former and leave the latter (which is straightforward) until our discussion of three-dimensional solids.

The nearly-free electron model demands the application of very elementary perturbation theory. It is possible that many chemists, although familiar with the variation method, may be less familiar with the perturbation method of approximation. For readers in this situation, consultation of an elementary textbook on quantum mechanics will be necessary although we give without proof in brief outline a reminder of the basic elements of the first order theory.

Let us suppose that the Schrödinger equation for the non-degenerate state of some system is

$$H\phi = E\phi \tag{6.64}$$

and that we may express the Hamiltonian operator as the sum

$$H = H^0 + H' \tag{6.65}$$

where the effect of H' on the wave functions is small. Let us suppose further that the equation

$$H^0\psi = E^0\psi \tag{6.66}$$

is soluble subject to the same boundary conditions as equation 6.64. We may thus associate H^0 with the unperturbed system, H with the perturbed system and H' with the perturbation. If the perturbation is very small, then ϕ and ψ will differ very little (here we are focussing attention on a particular state) and according to first order perturbation theory, we may write

$$\begin{aligned} E &= \frac{\int \psi^* H \psi \, \mathrm{d}\tau}{\int \psi^* \psi \, \mathrm{d}\tau} \\ &= \frac{\int \psi^* (H^0 + H') \psi \, \mathrm{d}\tau}{\int \psi^* \psi \, \mathrm{d}\tau} \\ &= E^0 + \frac{\int \psi^* H' \psi \, \mathrm{d}\tau}{\int \psi^* \psi \, \mathrm{d}\tau} \\ &= E^0 + E' \end{aligned} \tag{6.67}$$

In other words, the first order perturbation energy E' is obtained by averaging the perturbation term over the *unperturbed* states of the system.

Let us now suppose that the unperturbed state ψ is doubly degenerate i.e.

$$H^0\psi_1 = E^0\psi_1 \tag{6.68}$$

and

$$H^0\psi_2 = E^0\psi_2 \tag{6.69}$$

Any linear combination

$$\psi = c_1\psi_1 + c_2\psi_2 \tag{6.70}$$

is also a solution, so that

$$H^0(c_1\psi_1 + c_2\psi_2) = E^0(c_1\psi_1 + c_2\psi_2) \tag{6.71}$$

Thus, adopting the same formalism as above we may assume that

$$(H^\circ + H')(c_1\psi_1 + c_2\psi_2) = (E^0 + E')(c_1\psi_1 + c_2\psi_2) \tag{6.72}$$

Combination of eqns 6.71 and 6.72 leads to

$$(H' - E')(c_1\psi_1 + c_2\psi_2) = 0 \tag{6.73}$$

from which we may obtain the two equations

$$\int \psi_1^*(H' - E')(c_1\psi_1 + c_2\psi_2)\, d\tau = 0 \tag{6.74}$$

and

$$\int \psi_2^*(H' - E')(c_1\psi_1 + c_2\psi_2)\, d\tau = 0 \tag{6.75}$$

If we write the *matrix elements** as H'_{ij} eqns 6.74 and 6.75 reduce to

$$c_1H'_{11} + c_2H'_{12} - E'c_1 \int \psi_1^*\psi_1\, d\tau - E'c_2 \int \psi_1^*\psi_2\, d\tau = 0 \tag{6.76}$$

and

$$c_1H'_{21} + c_2H'_{22} - E'c_1 \int \psi_2^*\psi_1\, d\tau - E'c_2 \int \psi_2^*\psi_2\, d\tau = 0 \tag{6.77}$$

We assume for convenience that ψ_1 and ψ_2 are orthogonal and normalised, so that from eqns 6.76 and 6.77

$$c_1(H'_{11} - E') + c_2H'_{12} = 0 \tag{6.78}$$

* This term, which need not concern us particularly, comes from matrix mechanics and is merely $\int \psi_i^* H'\psi_j\, d\tau$ or in Dirac formalism $\langle\psi_i|H'|\psi_j\rangle$.

and

$$c_2(H'_{22} - E') + c_1 H_{21} = 0 \tag{6.79}$$

For the non trivial solutions of 6.78 and 6.79 ($c_1 = c_2 = 0$ is not permitted anyway) we must equate the *secular determinant* to zero

$$\begin{vmatrix} H'_{11} - E' & H'_{12} \\ H'_{21} & H'_{22} - E' \end{vmatrix} = 0 \tag{6.80}$$

Expansion of 6.80 will lead to the *secular equation* which will be quadratic in E' and in general will have two *distinct* roots. In other words, the perturbation may split the doubly degenerate unperturbed level into two distinct levels (on the other hand 6.80 might lead to two equal roots when $H'_{11} = H'_{22}$ and $H'_{12} = 0$, so that degeneracy need not *necessarily* be split by the first order effect).

Before we finally manage to consider the matrix elements of the potential energy we first need to prove for convenience that if the periodic potential $V(x)$ has periodicity a, then

$$\int_0^{L=Na} \exp(ikx)V(x)\,\mathrm{d}x = 0 \tag{6.81}$$

unless

$$k = \frac{2\pi n}{a} \qquad (n = 0, \pm 1, \pm 2\cdots)$$

The integral of eqn. 6.81 may be expressed in the form

$$\sum_{m=0}^{N-1} \int_{ma}^{(m+1)a} \exp(ikx)V(x)\,\mathrm{d}x \tag{6.82}$$

and by changing the variable to y such that $x = y + ma$ we have,

$$\int_{ma}^{(m+1)a} \exp(ikx)V(x)\,\mathrm{d}x = \int_0^a \exp[ik(y + ma)]V(y + ma)\,\mathrm{d}y$$

$$= \exp(ikma) \int_0^a \exp(iky)V(y)\,\mathrm{d}y \tag{6.83}$$

Hence every term in the summation 6.82 is identical, so instead we write

$$\sum_{m=0}^{N-1} \exp(ikma) \int_0^a \exp(ikx)V(x)\,\mathrm{d}x \tag{6.84}$$

Now

$$\sum_{m=0}^{N-1} \exp(ikma) = \frac{1 - \exp(iNka)}{1 - \exp(ika)} \text{ (see eqn. 3.48)}$$
$$= 0 \tag{6.85}$$

unless $\exp(ika) = 1$ (or $k = 2\pi n/a$) when the sum is N.

The value of this theorem may be seen by considering the matrix element

$$V_{kk'} = \langle k | V(x) | k' \rangle$$
$$= \int_0^L \psi_k^*(x) V(x) \psi_{k'}(x) \, dx$$
$$= \int_0^L \exp[i(k' - k)x] V(x) u_k^*(x) u_{k'}(x) \, dx \tag{6.86}$$

Since $V(x)u_k^*(x)u_{k'}(x)$ has the same periodicity at $V(x)$, eqn. 6.86 is of the same form as 6.81 so that we may write

$$V_{kk'} = 0 \text{ unless } k' - k = \frac{2\pi n}{a} \tag{6.87}$$

We may now calculate the energy gaps for a one-dimensional lattice, assuming that $V(x)$ is of sufficiently small magnitude that it may be treated as a small perturbation of the free electron gas. We recall eqn. 5.51, here repeated for convenience

$$\psi_k(x) = \left(\frac{1}{L}\right)^{1/2} \exp(ikx) \tag{6.88}$$

which we take as the unperturbed function, noting that ψ_k and ψ_{-k} are degenerate, both corresponding to the unperturbed energy

$$\varepsilon^0 = \frac{\hbar^2 k^2}{2m} \tag{6.89}$$

The functions are orthonormal so that we may write for the secular determinant (6.80)

$$\begin{vmatrix} V_{kk} - \varepsilon' & V_{k(-k)} \\ V_{(-k)k} & V_{(-k)(-k)} - \varepsilon' \end{vmatrix} = 0 \tag{6.90}$$

where the matrix elements are as follows

$$V_{kk} = V_{(-k)(-k)} = \frac{1}{L} \int_0^L V(x) \, dx \tag{6.91}$$

$$V_{k(-k)} = \frac{1}{L}\int_0^L \exp(-2ikx)V(x)\,\mathrm{d}x \tag{6.92}$$

$$V_{(-k)k} = \frac{1}{L}\int_0^L \exp(2ikx)V(x)\,\mathrm{d}x \tag{6.93}$$

Since V_{kk} is just the average of $V(x)$ over the lattice, we may conveniently choose our zero of energy so that this average is zero. This simplified secular equation is then

$$\begin{aligned}\varepsilon'^2 &= V_{k(-k)}V_{(-k)k}\\ &= V_{k(-k)}V^*_{k(-k)}\\ &= |V_{k(-k)}|^2\end{aligned} \tag{6.94}$$

From eqn. 6.87, we see that $V_{k(-k)} = 0$ unless $2k = 2\pi n/a$ or

$$k = \frac{\pi n}{a} \qquad (n = 0 \pm 1, \pm 2\cdots) \tag{6.95}$$

In other words, first order perturbation theory predicts that the small periodic potential $V(x)$ will not split the doubly degenerate levels except at $k = \pi n/a$, when two perturbed levels of energy

$$\frac{\hbar^2}{2m}\left(\frac{n\pi}{a}\right)^2 \pm |V_n| \tag{6.96}$$

(where V_n is the matrix element of the type 6.92) are expected.

V_n is in fact any one of the coefficients of the Fourier expansion of the potential $V(x)$

$$V(x) = \sum_{n=-\infty}^{\infty} V_n \exp(iGx) \tag{6.97}$$

where G is a reciprocal lattice number. Proving this is simple and is left to the interested reader, who might also consider second order effects.

6.3 The three-dimensional periodic potential

Bloch functions and band gaps

Having discussed the nature of Bloch functions in the one-dimensional periodic lattice we may assume the form of the analogous functions in three dimensions. The Schrödinger equation for an electron in the periodic potential defined by

$$V(\boldsymbol{r}) = V(\boldsymbol{r} + \boldsymbol{R}) \tag{6.98}$$

where $\boldsymbol{R}$ is a real lattice vector (eqn. 3.9), is

$$\left[-\frac{\hbar^2}{2m}\nabla^2 + V(\boldsymbol{r})\right]\psi = \varepsilon\psi \tag{6.99}$$

and subject to the periodic boundary conditions

$$\psi(\boldsymbol{r}) = \psi(\boldsymbol{r} + N_1\boldsymbol{a}_1) = \psi(\boldsymbol{r} + N_2\boldsymbol{a}_2) = \psi(\boldsymbol{r} + N_3\boldsymbol{a}_3) \tag{6.100}$$

The Bloch condition (see 6.25) may be written

$$\psi_{\boldsymbol{k}}(\boldsymbol{r} + \boldsymbol{R}) = \exp(i\boldsymbol{k}\cdot\boldsymbol{R})\psi_{\boldsymbol{k}}(\boldsymbol{r}) \tag{6.101}$$

and the Bloch function as

$$\psi_{\boldsymbol{k}}(\boldsymbol{r}) = u_{\boldsymbol{k}}(\boldsymbol{r})\exp(i\boldsymbol{k}\cdot\boldsymbol{r}) \tag{6.102}$$

where the modulation function $u_{\boldsymbol{k}}(\boldsymbol{r})$ has the lattice periodicity

$$u_{\boldsymbol{k}}(\boldsymbol{r}) = u_{\boldsymbol{k}}(\boldsymbol{r} + \boldsymbol{R}) \tag{6.103}$$

Hence from 6.100 and 6.101 it follows that

$$\boldsymbol{k}\cdot\boldsymbol{a}_1 = \frac{2\pi n_1}{N_1} \tag{6.104}$$

$$\boldsymbol{k}\cdot\boldsymbol{a}_2 = \frac{2\pi n_2}{N_2} \tag{6.105}$$

$$\boldsymbol{k}\cdot\boldsymbol{a}_3 = \frac{2\pi n_3}{N_3} \tag{6.106}$$

where $n_1 n_2$ and n_3 are integers.

It will be noted that eqns 6.104–6.106 are of the same form as the Laue equations (3.51–3.53), so by a similar sequence of steps, we may infer that

$$\begin{aligned}\boldsymbol{k} &= \frac{n_1}{N_1}\boldsymbol{b}_1 + \frac{n_2}{N_2}\boldsymbol{b}_2 + \frac{n_3}{N_3}\boldsymbol{b}_3 \\ &= \boldsymbol{G}\end{aligned} \tag{6.107}$$

Recalling that $\exp(i\boldsymbol{G}\cdot\boldsymbol{R}) = 1$ (eqn. 3.17) it will be noted that $\exp(i\boldsymbol{G}\cdot\boldsymbol{r})$ has the periodicity of the lattice so that the Bloch functions are periodic with the periodicity of the reciprocal lattice. Thus in a way discussed in the one-dimensional case, in the reduced scheme we may conveniently limit $\boldsymbol{k}$-vectors to the first Brillouin zone whose boundary faces are defined by eqn. 3.26 which we repeat for reference

$$2\boldsymbol{k}\cdot\boldsymbol{G} = G^2 \tag{6.108}$$

where the vector $\boldsymbol{G}$ is the shortest of the reciprocal lattice.

In the one-dimensional lattice we established that band gaps occur at energies corresponding to the electron wave numbers $k = n\pi/a$. We now wish to obtain analogous information in three dimensions. With the nearly-free electron approximation still fairly fresh in our minds, let us apply the general arguments to the three-dimensional periodic lattice. We will draw analogies without proof, assuming that the serious student will wish to confirm independently the acceptability of the parallel development.

We take as our unperturbed function the free electron wave function (5.62). Although we have a multiple-degeneracy problem and the secular determinant is of higher order, the same criterion, that energy gaps can only occur when some of the off-diagonal matrix elements are finite, still applies. In the one-dimensional case the condition

$$V_{kk'} = \langle k|V(r)|k'\rangle \neq 0 \tag{6.109}$$

where $k' = -k$, had to be satisfied. In the three-dimensional case, the similar condition is

$$V_{\boldsymbol{kk'}} = \langle \boldsymbol{k}|V(\boldsymbol{r})|\boldsymbol{k'}\rangle \neq 0 \tag{6.110}$$

where the *magnitudes* $k = k'$.

We may evaluate the matrix element $V_{\boldsymbol{kk'}}$ making use of arguments analogous to those presented in eqns 6.81–6.87:

$$V_{\boldsymbol{kk'}} = \left(\frac{1}{L}\right)^3 \int \exp[i(\boldsymbol{k'} - \boldsymbol{k})\cdot\boldsymbol{r}]V(\boldsymbol{r})\,\mathrm{d}\boldsymbol{r} \tag{6.111}$$

If we substitute

$$\boldsymbol{r} = \boldsymbol{r} + \boldsymbol{R} \tag{6.112}$$

eqn. 6.111 becomes

$$V_{\boldsymbol{kk'}} = \left(\frac{1}{L}\right)^3 \sum_{\boldsymbol{R}} \exp[i(\boldsymbol{k'} - \boldsymbol{k})\cdot\boldsymbol{R}] \int \exp[i(\boldsymbol{k'} - \boldsymbol{k})\cdot\boldsymbol{r}]V(\boldsymbol{r})\,\mathrm{d}\boldsymbol{r} \tag{6.113}$$

The triple summation

$$\sum_{\boldsymbol{R}} \exp[i(\boldsymbol{k'} - \boldsymbol{k})\cdot\boldsymbol{R}] \tag{6.114}$$

should be familiar from our discussion of diffraction in Chapter 3 (see eqn. 3.44) and we may write down immediately the condition that the matrix element $V_{\boldsymbol{kk'}}$ will be zero unless

$$\boldsymbol{k'} - \boldsymbol{k} = \boldsymbol{G} \tag{6.115}$$

or since $k' = k$

$$2\boldsymbol{k}\cdot\boldsymbol{G} = G^2 \tag{6.116}$$

This beautiful result indicates that energy gaps will only occur at wave vectors defined by 6.116 or by reference to 6.108 at wave vectors defined by the Brillouin zone. Further, eqn. 6.116 is identically the diffraction condition obtained in our discussion of the Ewald sphere construction (see 3.61); our interpretation of energy bands in terms of Bragg reflexion in the one-dimensional lattice clearly obtains in three dimensions.

Brillouin zones in three dimensions

Our discussions of Chapter 3 should have been sufficient for the construction of the Brillouin zones of any three-dimensional lattice. However, the physical interpretation of energy gaps at zone boundaries suggests an alternative approach which is worthy of consideration.

For a structure with a basis, it is obvious that the lattice vectors of 6.114 must be further expanded as in 3.64 to include products with the structure factor $F(l_1 l_2 l_3)$. Hence if the structure factor is zero, (systematic absences) there is no energy discontinuity across the corresponding plane in $\boldsymbol{k}$-space. As example of this approach we refer to structures with cubic axes.

It may be readily confirmed that the reciprocal lattice vectors of the simple cubic lattice are defined by

$$\boldsymbol{G} = \frac{2\pi}{a}(l_1 \hat{\boldsymbol{x}}_1 + l_2 \hat{\boldsymbol{x}}_2 + l_3 \hat{\boldsymbol{x}}_3) \qquad (6.117)$$

so that the perpendicular distance of Wigner–Seitz planes of the unit cell represented by $(l_1 l_2 l_3)$ from the origin is given by

$$P(l_1 l_2 l_3) = \frac{G}{2}$$

$$= \frac{\pi}{a}(l_1{}^2 + l_2{}^2 + l_3{}^2)^{1/2} \qquad (6.118)$$

We recall for the *bcc* structure that $F(l_1 l_2 l_3)$ is zero when $(l_1 + l_2 + l_3)$ is an odd integer and is 2 when $(l_1 + l_2 + l_3)$ is an even integer. Hence the first planes in reciprocal lattice space corresponding to an energy gap are of the type $\{110\}$. Thus the first Brillouin zone is bounded by twelve planes at a distance of $\sqrt{2}\pi/a$ from the origin. We may treat similarly the *fcc* lattice, and it should be confirmed that since $F(111) = F(200) = 4$ while $F(100) = F(110) = 0$, the first Brillouin zone is bounded by eight $\{111\}$ planes and six $\{200\}$ planes at respective distances $\sqrt{3}\pi/a$ and $2\pi/a$ from the origin. For further experience of this approach to the construction of Brillouin zones, the reader may carry out a similar treatment on the diamond structure and some structures based on the hexagonal lattice (in case of confusion Mott and Jones should be consulted).

The tight binding approximation

This approach, with which we will deal only in outline, should be particularly appealing to chemists because it is really the familiar L.C.A.O. method. It was used originally by Bloch when he suggested that the linear combination of atomic orbitals derived from the atoms of a crystalline lattice should provide an acceptable approximate solution to the wave equation.

Let us first consider the free atom wave functions. We assume, as usual, a one-electron approximation, so that the potential of an electron in the atom is made up of a contribution from the nuclear field and a contribution from the average field of all of the other electrons. We designate this potential $V_0(\boldsymbol{S})$ where $\boldsymbol{S}$ is a position vector with respect to that atom. The ground state wave function of the free atom is thus

$$H_0\psi_0(\boldsymbol{S}) = \varepsilon_0\psi_0(\boldsymbol{S}) \tag{6.119}$$

where

$$H_0 = -\frac{\hbar^2}{2m}\nabla^2 + V_0(\boldsymbol{S}) \tag{6.120}$$

We will assume that the wave functions are normalised and ε_0 is non-degenerate. Now let us suppose that a number of identical atoms are brought together to make up a crystal by occupying lattice sites. The potential energy term is now periodic with the periodicity of the lattice. If we take an atom of the lattice as our origin, the position of any other atom may be described by the lattice vector $\boldsymbol{R}_i$. In the tight binding approach, it is assumed that an electron (at *P* in Fig. 6.9) which is in the vicinity of a particular atom *i* is only influenced by that atom; i.e. it is unaffected by the other atoms of the lattice. Thus we may write the wave function as

$$\psi(\boldsymbol{S}) = \psi_0(\boldsymbol{r} - \boldsymbol{R}_i) \tag{6.121}$$

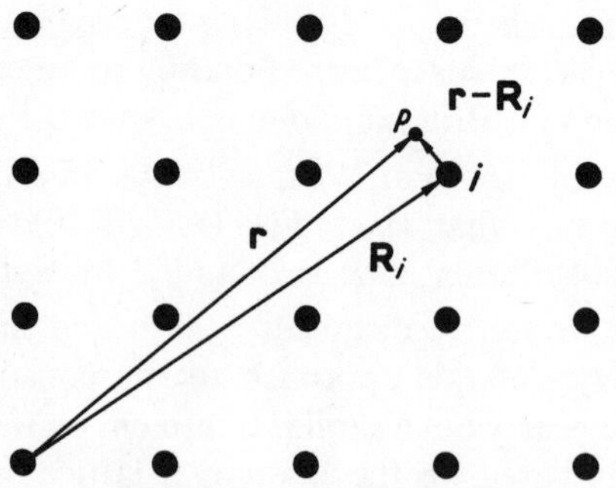

FIG. 6.9 An electron in the tight binding approach.

so that, in general, the crystal wave function may be written as a linear combination of the form

$$\psi(\boldsymbol{r}) = \sum_i c_i \psi_0(\boldsymbol{r} - \boldsymbol{R}_i) \tag{6.122}$$

The summation being over all atoms of the crystal, which for simplicity we assume is infinite. Since we are considering an electron in a periodic potential, $\psi(\boldsymbol{r})$ must be a Bloch function, so that it may be written in the following form by replacing the coefficients c_i.

$$\psi_{\boldsymbol{k}}(\boldsymbol{r}) = \sum_i \exp(i\boldsymbol{k} \cdot \boldsymbol{R}_i)\psi_0(\boldsymbol{r} - \boldsymbol{R}_i) \tag{6.123}$$

We may demonstrate that 6.123 is of the Bloch form by applying a transformation corresponding to a lattice vector $\boldsymbol{R}_j$ to give

$$\begin{aligned}
\psi_{\boldsymbol{k}}(\boldsymbol{r} + \boldsymbol{R}_j) &= \sum_i \exp(i\boldsymbol{k} \cdot \boldsymbol{R}_i)\psi_0(\boldsymbol{r} - \boldsymbol{R}_i + \boldsymbol{R}_j) \\
&= \exp(i\boldsymbol{k} \cdot \boldsymbol{R}_j) \sum_i \exp[i\boldsymbol{k} \cdot (\boldsymbol{R}_i - \boldsymbol{R}_j)]\psi_0[\boldsymbol{r} - (\boldsymbol{R}_i - \boldsymbol{R}_j)] \\
&= \exp(i\boldsymbol{k} \cdot \boldsymbol{R}_j)\psi_{\boldsymbol{k}}(\boldsymbol{r})
\end{aligned} \tag{6.124}$$

Thus the Bloch condition (6.101) is satisfied.

$\psi_{\boldsymbol{k}}(\boldsymbol{r})$ (eqn. 6.123) is a solution of the Schrödinger equation for the whole crystal:

$$H\psi_{\boldsymbol{k}} = \varepsilon\psi_{\boldsymbol{k}} \tag{6.125}$$

The crystal Hamiltonian H may be written as the sum

$$H = H_0 + H_1 \tag{6.126}$$

where

$$H_0 = -\frac{\hbar^2}{2m}\nabla^2 + V_0(\boldsymbol{r} - \boldsymbol{R}_i) \tag{6.127}$$

and

$$H_1 = V(\boldsymbol{r}) - V_0(\boldsymbol{r} - \boldsymbol{R}_i) \tag{6.128}$$

We now let H_0 operate on the function $\psi_{\boldsymbol{k}}(\boldsymbol{r})$

Hence

$$\begin{aligned}
H_0\psi_{\boldsymbol{k}}(\boldsymbol{r}) &= \sum_i \exp(i\boldsymbol{k} \cdot \boldsymbol{R}_i)H_0\psi_0(\boldsymbol{r} - \boldsymbol{R}_i) \\
&= \varepsilon_0 \sum_i \exp(i\boldsymbol{k} \cdot \boldsymbol{R}_i)\psi_0(\boldsymbol{r} - \boldsymbol{R}_i)
\end{aligned}$$

(from 6.119)

$$= \varepsilon_0\psi_{\boldsymbol{k}}(\boldsymbol{r}) \tag{6.129}$$

(from 6.123).

The energy of the electron in the crystal $\varepsilon_{\boldsymbol{k}}$ may be found by evaluating

$$\varepsilon_{\boldsymbol{k}} = \frac{\int \psi_{\boldsymbol{k}}^*(\boldsymbol{r})(H_0 + H_1)\psi_{\boldsymbol{k}}(\boldsymbol{r})\,\mathrm{d}\tau}{\int \psi_{\boldsymbol{k}}^*(\boldsymbol{r})\psi_{\boldsymbol{k}}(\boldsymbol{r})\,\mathrm{d}\tau} \tag{6.130}$$

Or by use of 6.123

$$\varepsilon_{\boldsymbol{k}} = \varepsilon_0 + \frac{\int \psi_{\boldsymbol{k}}^*(\boldsymbol{r}) \sum_i \exp(i\boldsymbol{k}\cdot\boldsymbol{R}_i)[V(\boldsymbol{r}) - V_0(\boldsymbol{r} - \boldsymbol{R}_i)]\psi_0(\boldsymbol{r} - \boldsymbol{R}_i)\,\mathrm{d}\tau}{\int \psi_{\boldsymbol{k}}^*(\boldsymbol{r})\psi_{\boldsymbol{k}}(\boldsymbol{r})\,\mathrm{d}\tau} \tag{6.131}$$

The numerator of eqn. 6.131 may be written as the double summation if substitution for $\psi_{\boldsymbol{k}}^*(\boldsymbol{r})$ is made from eqn. 6.123

$$\sum_i \sum_j \exp[i\boldsymbol{k}(\boldsymbol{R}_i - \boldsymbol{R}_j)] \int \psi_0^*(\boldsymbol{r} - \boldsymbol{R}_j)[V(\boldsymbol{r}) - V_0(\boldsymbol{r} - \boldsymbol{R}_i)]\psi_0(\boldsymbol{r} - \boldsymbol{R}_i)\,\mathrm{d}\tau$$

If by neglecting overlap at adjacent lattice sites we assume that

$$\int \psi_{\boldsymbol{k}}^*(\boldsymbol{r} - \boldsymbol{R}_i)\psi_{\boldsymbol{k}}(\boldsymbol{r} - \boldsymbol{R}_j)\,\mathrm{d}\tau = \delta_{ij} \tag{6.132}$$

and that

$$\int \psi_{\boldsymbol{k}}^*(\boldsymbol{r})\psi_{\boldsymbol{k}}(\boldsymbol{r}) = N \tag{6.133}$$

where N is the total number of electrons in the crystal; it follows that

$$\varepsilon_{\boldsymbol{k}} = \varepsilon_0 + \frac{1}{N}\sum_i \sum_j \exp[i\boldsymbol{k}\cdot(\boldsymbol{R}_i - \boldsymbol{R}_j)] \int \psi_0^*(\boldsymbol{r} - \boldsymbol{R}_j)$$
$$\times\ [V(\boldsymbol{r}) - V_0(\boldsymbol{r} - \boldsymbol{R}_i)]\psi_0(\boldsymbol{r} - \boldsymbol{R}_i)\,\mathrm{d}\tau \tag{6.134}$$

We expect every term in the summation $i = 0$ to $i = N - 1$ to contain N identical terms, which may be most easily evaluated by substituting $i = 0$. Hence

$$\varepsilon_{\boldsymbol{k}} = \varepsilon_0 + \sum_j \exp(-i\boldsymbol{k}\cdot\boldsymbol{R}_j) \int \psi_0^*(\boldsymbol{r} - \boldsymbol{R}_j)[V(\boldsymbol{r}) - V_0(\boldsymbol{r})]\psi_0(\boldsymbol{r})\,\mathrm{d}\tau \tag{6.135}$$

If we now make the further approximation that ψ_0 is spherically symmetric, all nearest neighbour contributions may be assumed to be identical. We may neglect other than nearest neighbour interactions. We need only consider therefore two contributions to the summation of 6.135.

For $j = 0$, we let

$$\int \psi_0^*(\boldsymbol{r})[V(\boldsymbol{r}) - V_0(\boldsymbol{r})]\psi_0(\boldsymbol{r}) = -\alpha \tag{6.136}$$

and for nearest neighbour atoms

$$\int \psi_0^*(\boldsymbol{r} - \boldsymbol{R}_j)[V(\boldsymbol{r}) - V_0(\boldsymbol{r})]\psi_0(\boldsymbol{r}) = -\gamma \tag{6.137}$$

Hence

$$\varepsilon_{\boldsymbol{k}} = \varepsilon_0 - \alpha - \gamma \sum_j \exp(i\boldsymbol{k} \cdot \boldsymbol{R}_j) \tag{6.138}$$

it being *understood* that the summation is to be carried out over nearest neighbours only.

We may observe from 6.138 that the energy of an electron consists of a constant term $\varepsilon_0 - \alpha$ together with a term dependent on $\boldsymbol{k}$. It is the latter term which transforms the discrete levels of the atom into the energy bands of the crystal. Let us first demonstrate an elementary application of the method to the simple cubic lattice. In this lattice a chosen atom at the origin has six nearest neighbours located at the vector components of $\boldsymbol{R}_j$ given by

$$\boldsymbol{R}_j = (\pm a, 0, 0); (0, \pm a, 0); (0, 0, \pm a)$$

Evaluation of the summation of 6.138 and substitution lead to

$$\varepsilon_{\boldsymbol{k}} = \varepsilon_0 - \alpha - 2\gamma(\cos k_1 a + \cos k_2 a + \cos k_3 a) \tag{6.139}$$

and we immediately observe that the $\boldsymbol{k}$-dependent part of $\varepsilon_{\boldsymbol{k}}$ is periodic with periodicity of $2\pi/a$ in the three k-components which we may restrict to the first Brillouin zone defined by

$$-\frac{\pi}{a} \leq k_1 \leq \frac{\pi}{a}$$

$$-\frac{\pi}{a} \leq k_2 \leq \frac{\pi}{a}$$

$$-\frac{\pi}{a} \leq k_3 \leq \frac{\pi}{a}$$

Further, since the cosine terms are limited to the range -1 to $+1$, the maximum band width is 12γ. The integral γ, and therefore the band width, will be greater the more the wave functions ψ_0 overlap. We may note that at the Brillouin zone boundary planes, the normal derivative $\mathrm{d}\varepsilon/\mathrm{d}k = 0$ (easily seen from the differential of 6.139). This is a further illustration than at the zone boundaries the wave functions are no longer travelling waves, but stationary waves which cannot propagate through the lattice. In other words, at the zone boundaries

$$\boldsymbol{v} = \frac{1}{\hbar}\,\mathrm{grad}_{\boldsymbol{k}}\,\varepsilon_{\boldsymbol{k}} \tag{6.140}$$

$$= 0$$

For brevity we will not deal with other types of lattice, but leave the interested reader to consider say *fcc* and *bcc* lattices.

6.4 The effective mass of electrons and holes

Let us consider the consequences of the assumption that the Bloch electrons obey the classical relationship between force and energy

$$\frac{\mathrm{d}\varepsilon_{\boldsymbol{k}}}{\mathrm{d}t}\,\boldsymbol{k} = \boldsymbol{F}\cdot\boldsymbol{v} \tag{6.141}$$

where $\boldsymbol{F}$ is an applied force and $\boldsymbol{v}$ = velocity. But we know from the group velocity equation (6.140) that

$$\frac{\mathrm{d}\boldsymbol{v}}{\mathrm{d}t} = \frac{1}{\hbar}\,\mathrm{grad}_{\boldsymbol{k}}\left(\frac{\mathrm{d}\varepsilon_{\boldsymbol{k}}}{\mathrm{d}t}\right) \tag{6.142}$$

Hence by substituting 6.140 and 6.141 we obtain

$$\frac{\mathrm{d}\boldsymbol{v}}{\mathrm{d}t} = \frac{1}{\hbar}\,\mathrm{grad}_{\boldsymbol{k}}(\boldsymbol{F}\cdot\boldsymbol{v})$$

$$= \frac{1}{\hbar^2}\,\boldsymbol{F}\cdot(\mathrm{grad}_{\boldsymbol{k}}\,\mathrm{grad}_{\boldsymbol{k}}\,\varepsilon_{\boldsymbol{k}}) \tag{6.143}$$

where $\mathrm{grad}_{\boldsymbol{k}}\,\mathrm{grad}_{\boldsymbol{k}}\,\varepsilon_{\boldsymbol{k}}$ is the second rank tensor

$$\begin{bmatrix} \dfrac{\partial^2\varepsilon}{\partial k_1{}^2} & \dfrac{\partial^2\varepsilon}{\partial k_2\,\partial k_1} & \dfrac{\partial^2\varepsilon}{\partial k_3\,\partial k_1} \\ \dfrac{\partial^2\varepsilon}{\partial k_1\,\partial k_2} & \dfrac{\partial^2\varepsilon}{\partial k_2{}^2} & \dfrac{\partial^2\varepsilon}{\partial k_3\,\partial k_2} \\ \dfrac{\partial^2\varepsilon}{\partial k_1\,\partial k_3} & \dfrac{\partial^2\varepsilon}{\partial k_2\,\partial k_3} & \dfrac{\partial^2\varepsilon}{\partial k_3{}^2} \end{bmatrix} \tag{6.144}$$

Equation 6.143 is analogous to the classical equation

$$\frac{\mathrm{d}\boldsymbol{v}}{\mathrm{d}t} = \frac{\boldsymbol{F}}{m} \tag{6.145}$$

and comparison shows that the quantity replacing the m is a second rank tensor, so that the acceleration need not be in the same direction as the applied force. This quantity is normally called the effective mass m^* defined by the equation

$$\frac{1}{m^*} = \frac{1}{\hbar^2} \operatorname{grad}_{\boldsymbol{k}} \operatorname{grad}_{\boldsymbol{k}} \varepsilon_{\boldsymbol{k}} \tag{6.146}$$

A special case of considerable importance occurs when ε is a general quadratic function of $\boldsymbol{k}$ (e.g. not too far from the bottom of a band, so that the surfaces of constant energy in $\boldsymbol{k}$-space are ellipsoids. If we choose our coordinate system according to the direction of the principal axes of the ellipsoid, we may define three principal effective masses according to the diagonal components of 6.144.

$$m_i^* = \frac{\hbar^2}{\partial^2 \varepsilon / \partial k_i^{\,2}} \; (i = 1, 2, 3) \tag{6.147}$$

so

$$\varepsilon_{\boldsymbol{k}} = \frac{\hbar^2}{2} \left(\frac{k_1^{\,2}}{m_1^*} + \frac{k_2^{\,2}}{m_2^*} + \frac{k_3^{\,2}}{m_3^*} \right) \tag{6.148}$$

The simplest possible situation, and one which is frequently assumed, is when an isotropic effective mass (a scalar) is used. Then

$$\varepsilon_{\mathbf{k}} = \frac{\hbar^2}{2m^*} (k_1^{\,2} + k_2^{\,2} + k_3^{\,2}) \tag{6.149}$$

Equations 6.146–6.149 clearly demonstrate the importance of $\varepsilon - \boldsymbol{k}$ relationships in determining effective mass. For the one-dimensional lattice, we represent the important relationships for an electron, ε *versus* k, $\mathrm{d}\varepsilon/\mathrm{d}k$ *versus* k, $\mathrm{d}^2\varepsilon/\mathrm{d}k^2$ *versus* k and m^* *versus* k in qualitative form in Fig. 6.10. We may immediately conclude that the effective mass of an electron is *positive* in the lower part of the band, but is *negative* close to the zone boundary. An electron of negative effective mass implies that its acceleration will occur in a direction opposite from that of the applied force. The type of force which would normally be considered is electrostatic or magnetic, so that an electron of *negative* effective mass would appear to have *positive* charge.

Vacant states in a band are commonly referred to as *hole states* and we will consider for simplicity a single hole near the top of an otherwise filled band (as might be imagined to arise in a semiconductor). The properties which may be associated with the hole are determined by the totality of electrons in the band. In one-dimensional language, if the electron is missing

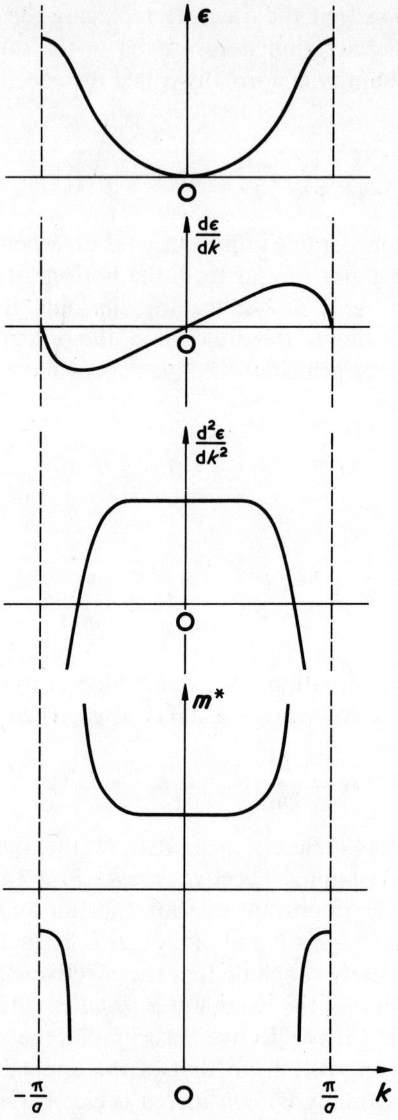

FIG. 6.10 Effective mass as a function of k in the one-dimensional lattice.

from the state k_1, the total wave number of the electrons in the band will be $-k_1$, which will be associated with the hole. Since the one-dimensional force law for an electron is

$$F = \hbar \frac{dk}{dt} \tag{6.150}$$

(this may be readily shown from the one-dimensional forms of 6.140 and 6.141) it follows that if the $-k_1$ is associated with the hole, it will behave under the influence of an electric or magnetic field as if it had a *positive charge* and *positive effective mass.* Thus in a semiconductor, both electrons and holes may be considered to act as carriers.

Admittedly our discussion of this topic has been rather sketchy, since other than recognising the useful concept of effective mass and the existence of hole states (the co-existence of electrons and hole is important in semiconductors) we will not refer again to it in detail. Several important questions have been left unanswered (for example, according to our discussion there appears to be no difference between a single electron and a single hole near the top of a band), but it is hoped that anxious readers will consult Kittel where will be found also a physical interpetation of effective mass.

6.5 Fermi surfaces and density of states in a Brillouin zone

The density of states formalism played a dominant role in our discussion of free electrons and now we make some qualitative enquiries about the density of states in a Brillouin zone. Let us return to the simple cubic lattice using the tight binding results of eqn. 6.139 and consider small values of k; we may expand the cosines, retaining only the second terms

$$\varepsilon_{\boldsymbol{k}} \simeq \varepsilon_0 - \alpha - 2\gamma\left[\left(1 - \frac{k_1{}^2a^2}{2}\right) + \left(1 - \frac{k_2{}^2a^2}{2}\right) + \left(1 - \frac{k_3{}^2a^2}{2}\right)\right]$$

$$= \varepsilon_0 - \alpha - 6\gamma + \gamma a^2(k_1{}^2 + k_2{}^2 + k_3{}^2) \tag{6.151}$$

In other words, near the bottom of the band we have a situation where $\varepsilon_{\boldsymbol{k}} \propto k^2$ as for free electrons. Constant energy surfaces are thus spheres in $\boldsymbol{k}$-space. Hence in this region, electrons may be considered as *free*, but with an effective mass determined by

$$\frac{\partial^2 \varepsilon}{\partial k^2} = 2\gamma a^2$$

so that

$$m^* = \frac{\hbar^2}{2\gamma a^2} \tag{6.152}$$

The filled band ($\varepsilon_{\boldsymbol{k}}$ a maximum) corresponds to a $\boldsymbol{k}$-value of $(\pm\pi/a, \pm\pi/a, \pm\pi/a)$ i.e. the corners of the Brillouin zone. Here again we may use an expansion but this time from cosines of the type $\cos k_1 a = \cos(k_1' a \pm \pi)$, where the new component $k_1' = k_1 \pm \pi/a$ is measured relative to the corner point. For small values of k' (the band almost full)

$$\varepsilon_{\boldsymbol{k}'} = \varepsilon_0 - \alpha + 2\gamma(\cos k_1' a + \cos k_2' a + \cos k_3' a) \tag{6.153}$$

may be expanded as was done previously to give

$$\varepsilon_{\boldsymbol{k}'} \simeq \varepsilon_0 - \alpha + 6\gamma - \gamma a^2(k_1'^2 + k_2'^2 + k_3'^2) \tag{6.154}$$

So again the constant energy surfaces are spheres, centred on the corners of the first Brillouin zone (see Fig. 6.11). It is not too difficult to tackle the problem of constant energy surfaces for other cubic structures, and for those readers who have derived the expressions for $\varepsilon_{\boldsymbol{k}}$ for the *fcc* and *bcc* lattices (an exercise previously suggested) it should prove worthwhile.

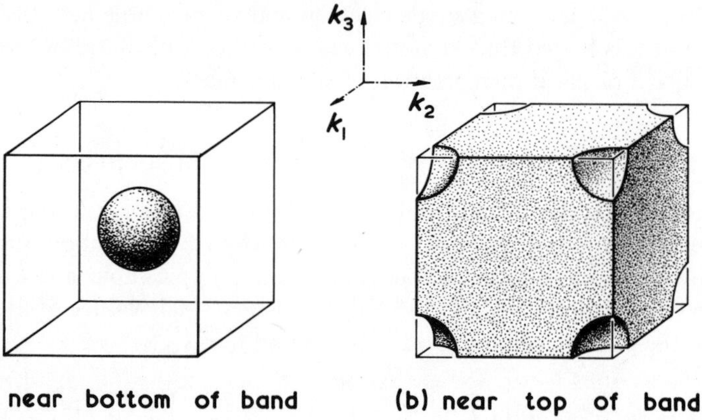

FIG. 6.11 Constant energy surfaces for the *sc* lattice from the tight binding approximation.

It is a convenient practice to represent the constant energy surfaces of the Brillouin zones in terms of a contour diagram. For example, contour curves for a two-dimensional square lattice are shown in Fig. 6.12 according to the tight-binding and nearly free electron models. But a related representation involving density of states functions is perhaps more useful. We will consider in outline the expected form of this representation for the simple cubic lattice. We may crudely assume, in the first instance, that spherical energy surfaces are generated about $\boldsymbol{k} = 0$ until the Fermi sphere touches the zone boundaries at $k_1 = k_2 = k_3 = \pm\pi/a$; then spherical surfaces are generated about the corners of the first Brillouin zone. The density of states functions (with

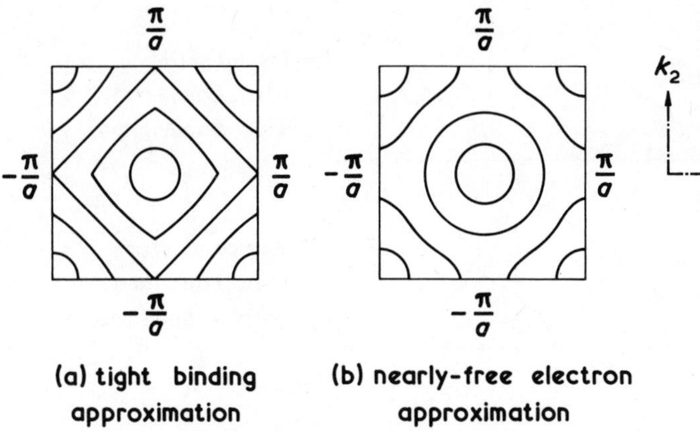

FIG. 6.12 Electron energy contours for the *sc* lattice.

appropriate choice of effective mass) would be those used in the free electron approximation (see eqn. 5.73) leading to the very simple $\mathcal{N}(\varepsilon)$ versus ε relationship presented in Fig. 6.13(a). However, as was implied in Figs 6.11 and 6.12 and as may be seen from our analysis of $\varepsilon_{\boldsymbol{k}}$ for the cubic lattice, spherical surfaces are only really acceptable for $\boldsymbol{k}$-values close to the bottom and top of the band. In fact if we construct an $\mathcal{N}(\varepsilon)$ curve on the basis of the tight binding approximation, the result shown in Fig. 6.13(b) is obtained. The parabolicity of the curve at each end will be noted. Finally in Fig. 6.13(c) is shown a diagram, based on a more realistic model, which may be considered qualitatively as a compromise between (a) and (b).

This representation is commonly used in presenting in an elementary way the contrasting band structures of metals, insulators and semiconductors. Fig. 6.14 is self-explanatory, except that we introduce for the first time the idea of overlapping bands. In our frequent use of a one-dimensional language this possibility cannot arise, but as soon as we move to two or three dimensions the feasibility of the smallest $\boldsymbol{k}$-vector necessary for an electron to occupy the second Brillouin zone being (in the extended zone representation)

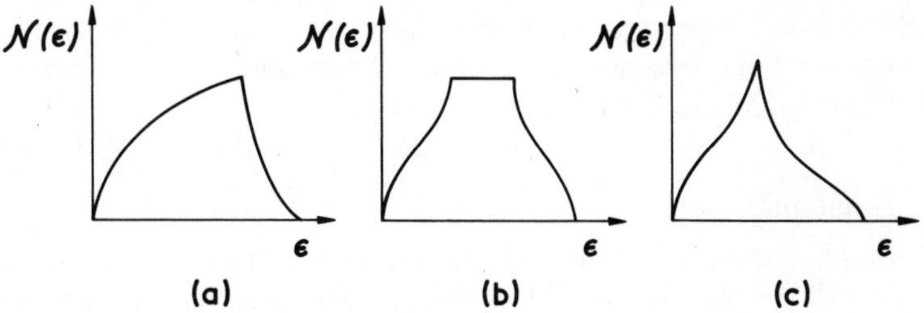

FIG. 6.13 Density of states functions for the simple cubic lattice.

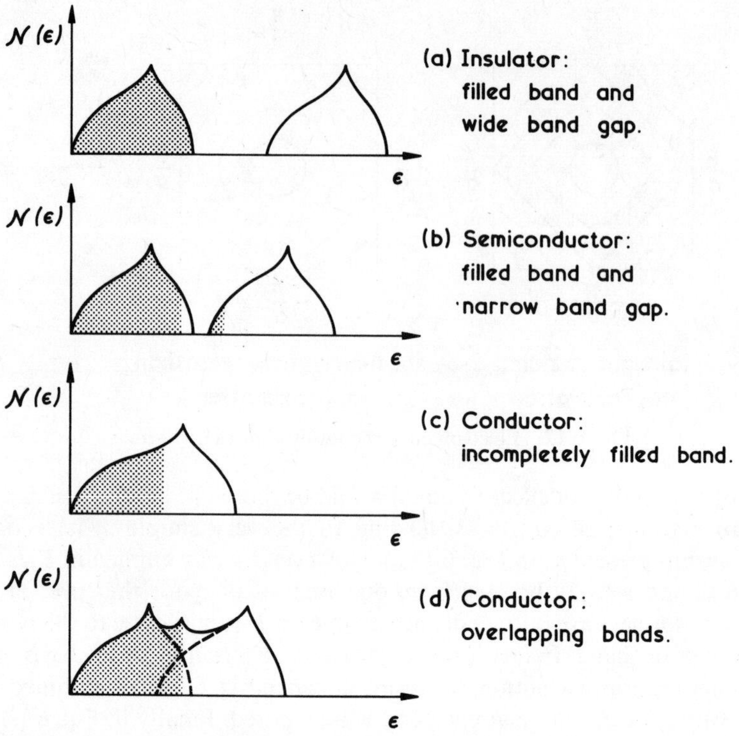

FIG. 6.14 Band structures of insulators, semiconductors and conductors.

less than that for the occupation of levels close to the boundary of the first zone, naturally arises. An example of the behaviour may be cited in the alkaline earth metals (note that we mentioned this apparent anomaly when discussing the one-dimensional case).

It is hoped that by this stage, the reader will have a reasonably well-developed idea of the band approach to solid state properties, although some enquiry into the practical problem of the experimental study of band structures might be desirable. It is certain, however, that we have sufficient background and experience to begin discussing that very important and interesting class of solids, semiconductors. To most chemists, semiconductors are sufficiently alien to merit separate treatment.

6.6 Semiconductors

It is a frequent misconception that the history of semiconductors is limited to the last thirty years or so; in fact Michael Faraday and probably others noted that certain substances have a negative temperature coefficient of resistance, one of the characteristics of semiconductors. Perhaps the first major advances were made by Smith in 1873, who observed photoconductivity in selenium and by Braun, in 1874, who observed rectification using lead sulphide and other materials. Subsequently, semiconductors emerged as a defined class of materials with the following characteristics.

(1) A resistivity in the approximate range 10^{-3}–10^{6} Ω cm.
(2) A negative temperature coefficient of resistance.
(3) A high thermoelectric power.
(4) Non-ohmic behaviour.
(5) Photosensitivity.

In 1879 a very important paper by E. H. Hall[6] describing the development of a transverse potential drop across a conductor carrying an electric current in a magnetic field, was published. The observation of the *Hall effect* was to play a significant role in the study of current carriers in semiconductors. For example, it is observed that the *number* of current carriers increases with temperature, (in contrast with the behaviour of metals) accounting for the negative coefficient of resistance.

It is now known, and hinted at in earlier publications, that the properties of semiconducting materials are dramatically sensitive to the presence of small amounts of impurity. Indeed, with the limited techniques available, it is surprising that any consistent observations on a semiconductor such as silicon were possible. However, Hall coefficients were measured and besides the negative values expected for an electronic carrier, positive coefficients were observed, only to be understood at a later stage.

Some common semiconducting materials such as selenium were used as rectifiers on a limited scale at the end of the last century, but were not common until thirty years later when a copper oxide device was introduced as a low power rectifier. In this same period, high frequency rectification using a crystal of galena (lead sulphide) and a fine metal contact was recognised and the early days of commercial broadcasting saw the development of the art of manipulating the 'cat's whisker' to a state of finesse.

Although the crystal detector was replaced by the thermionic valve it emerged in very high frequency devices in radar development during the Second World War. Recognition of the commercial possibilities of semiconducting devices led to development work of febrile intensity, and finally the transistor was invented in 1949. The reader will be familiar with the consequences of this invention!

Intrinsic semiconductivity

A pure semiconducting material such as silicon at low temperature would exhibit a low conductivity. In contrast, a higher conductivity would result from the addition of very small amounts of an impurity such as phosphorus (this process is referred to as *doping*). We label the pure material an *intrinsic* semiconductor (since the properties result from the intrinsic electronic structure) and the doped material an *extrinsic or impurity* semiconductor (since its properties are dependent on some extrinsic agent). Initially we shall discuss the simpler case of intrinsic behaviour.

Adjusting our language to that appropriate to a one-dimension representation, it is common to refer to a *band gap*, whose magnitude, which we will always refer to as ε_g, is of central importance in discussing semiconductor properties. It may be of interest to list some values of ε_g for a range of materials (see Table 6.1). Let us describe briefly what we wish to achieve in our initial

TABLE 6.1 Band gaps of common semiconductors

Semiconductor crystal	ε_g(eV)
C (diamond)	5·33
Si	1·14
Ge	0·67
GaAs	1·4
GaP	2·25
CdS	2·42
ZnO	3·2
InSb	0·23
InAs	0·33
Te	0·33
PbS	0·26
PbSe	0·27

discussion. Assuming that we know the magnitude of the band gap, it is reasonable that we should wish to find out something about the electron and hole concentrations, since they determine, together with their respective mobilities, the conductivity of the bulk material at some chosen temperature. Since the thermal equilibrium of electrons is determined by the Fermi–Dirac function, a required knowledge of electron and hole concentrations amounts to a calculation of the position of μ (to which we will refer as the Fermi level).

We refer to Fig. 6.15; the lower band (filled at 0 K) is always called the *valence band* and the upper band is always called the *conductance band*. The density of states in these bands is respectively $\mathcal{N}_v(\varepsilon)$ and $\mathcal{N}_c(\varepsilon)$. The effective mass of electrons near the bottom of the conductance band is designated m_e^* and that of a hole near the top of the valence band m_h^*. Our chosen zero is at the top of the valence band and the Fermi level μ is arbitrarily placed in the band gap.

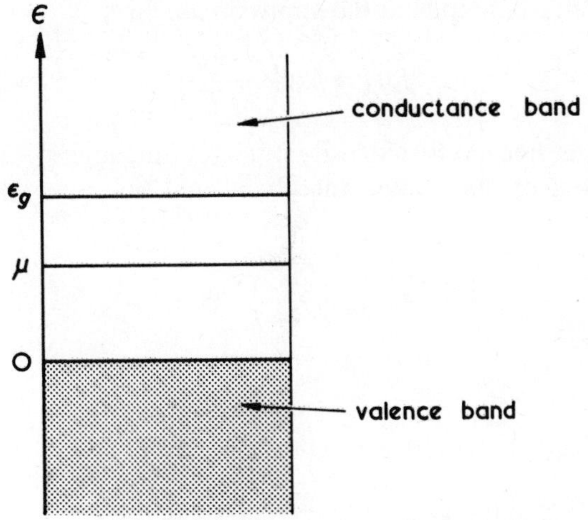

FIG. 6.15 The intrinsic semiconductor.

At 0 K, the total number of electrons, N which will just fill the valence band will be given by the expression

$$N = \int_{-\infty}^{0} \mathcal{N}_v(\varepsilon)\, d\varepsilon \tag{6.155}$$

but at any finite temperature, the occupation of electron states is determined by the Fermi–Dirac function. Thus we write

$$f_e(\varepsilon) = \frac{1}{\exp\left(\frac{\varepsilon - \mu}{k_B T}\right) + 1} \tag{6.156}$$

But since we may assume that *few* electrons occupy the conductance band, $f_e(\varepsilon) \ll 1$ so that the unity of the denominator may be neglected to give the classical limit

$$f_e(\varepsilon) \simeq \exp\left[-\frac{(\varepsilon - \mu)}{k_B T}\right] \tag{6.157}$$

For electrons in the valence band, we have in contrast $f_e(\varepsilon) \sim 1$, so that the function may be expanded to the approximate form

$$f_e(\varepsilon) \simeq 1 - \exp\left(\frac{\varepsilon - \mu}{k_B T}\right) \tag{6.158}$$

Thus the function for hole occupancy, which is obviously

$$f_h(\varepsilon) = 1 - f_e(\varepsilon) \tag{6.159}$$

may be alternatively written in the approximate form

$$f_h(\varepsilon) = \exp\left(\frac{\varepsilon - \mu}{k_B T}\right) \tag{6.160}$$

Hence, adopting free electron density of states functions (5.72) for electrons near the bottom of the conductance band and holes near the top of the valence band

$$\mathcal{N}_c(\varepsilon) = \frac{V}{2\pi^2}\left(\frac{2m_e^*}{\hbar^2}\right)^{3/2}(\varepsilon - \varepsilon_g)^{1/2} \tag{6.161}$$

and

$$\mathcal{N}_v(\varepsilon) = \frac{V}{2\pi^2}\left(\frac{2m_h^*}{\hbar^2}\right)^{3/2}(-\varepsilon)^{1/2} \tag{6.162}$$

so that from 6.157 and 6.160–6.162

$$n = \int_{\varepsilon g}^{\infty} \frac{1}{2\pi^2}\left(\frac{2m_e^*}{\hbar^2}\right)^{3/2}(\varepsilon - \varepsilon_g)^{1/2}\exp\left(\frac{\mu - \varepsilon}{k_B T}\right) d\varepsilon \tag{6.163}$$

and

$$p = \int_{-\infty}^{0} \frac{1}{2\pi^2}\left(\frac{2m_h^*}{\hbar^2}\right)^{3/2}(-\varepsilon)^{1/2}\exp\left(\frac{\varepsilon - \mu}{k_B T}\right) d\varepsilon \tag{6.164}$$

where n is the density of electrons in the conductance band and p is the density of holes in the valence band.

The integral in eqn. 6.163 may be evaluated readily if we conveniently change the variable by letting

$$\varepsilon = \varepsilon' + \varepsilon_g$$

so that 6.163 reduces to

$$n = \frac{1}{2\pi^2}\left(\frac{2m_e^*}{\hbar^2}\right)^{3/2}\exp\left(\frac{\mu - \varepsilon_g}{k_B T}\right)\int_0^{\infty}(\varepsilon')^{1/2}\exp\left(-\frac{\varepsilon'}{k_B T}\right) d\varepsilon' \tag{6.165}$$

We note the standard Fermi integral

$$\int_0^{\infty}\varepsilon^{1/2}\exp\left(-\frac{\varepsilon}{k_B T}\right) d\varepsilon = \frac{1}{2^{5/2}}\pi^{1/2}(2k_B T)^{3/2} \tag{6.166}$$

so that from 6.165

$$n = 2\left(\frac{m_e^* k_B T}{2\pi\hbar^2}\right)^{3/2}\exp\left(\frac{\mu - \varepsilon_g}{k_B T}\right) \tag{6.167}$$

From 6.164 we obtain in an analogous way

$$p = 2\left(\frac{m_h^* k_B T}{2\pi\hbar^2}\right)^{3/2} \exp\left(\frac{-\mu}{k_B T}\right) \tag{6.168}$$

The pre-exponential factors in 6.167 and 6.168 may be referred to respectively as the effective density of states $\mathcal{N}_c$ and $\mathcal{N}_v$. For an intrinsic semiconductor, the number of electrons in the conductance band must be equal to the number of holes in the valence band. So we have the condition that

$$n_i = p_i \tag{6.169}$$

Hence eqns 6.167 and 6.168 may be solved for μ to give

$$\mu = \frac{\varepsilon_g}{2} + 3/4 k_B T \ln\left(\frac{m_h^*}{m_e^*}\right) \tag{6.170}$$

Equation 6.170 is a statement of the well-known theorem that provided the effective masses of the electrons in the conductance band and the holes in the valence band are equal, the Fermi level is in the middle of the band gap. For many semiconductors, it is a good approximation. For some, however, (e.g. InSb for which $m_h^*/m_e^* = 20$ and $\varepsilon_g = 0.23$ eV) the Fermi level is shifted significantly from the mid-gap position at room temperature.

It will be observed that the product np is related to the band gap by the following eqn. (from 6.167 and 6.168)

$$np = 4\left(\frac{k_B T}{2\pi\hbar^2}\right)^3 (m_e^* m_h^*)^{3/2} \exp(-\varepsilon_g/k_B T) \tag{6.171}$$

which is independent of the Fermi level. It is, in fact, a statement of the law of mass action that np is a constant for a given material at a fixed temperature and may be used in discussing both intrinsic and extrinsic phenomena. Thus if n is increased by doping (see below), this must automatically mean a reduction in p.

The electrical conductivity of a material containing both electron and hole carriers will be

$$\sigma = ne\mu_e + pe\mu_h$$

where μ_e and μ_h are respectively the electron and hole mobilities (the over-use of the traditional symbol μ should cause no confusion) so that provided the major temperature dependence of σ results from the exponential dependence of n and p on reciprocal temperature (fortunately the mobilities are less temperature sensitive) we have immediately a method, in principle, of determining the energy gap ε_g. Alternatively an optical method may be used (see Kittel's discussion of direct and indirect band gaps).

Impurity semiconductivity

The doping of a semiconductor with a very low concentration of impurity atoms may alter its electronic properties dramatically, leading to what we have called extrinsic or impurity semiconductivity. Let us introduce this topic by means of the example of silicon or germanium doped with either phosphorus or arsenic. If we assume that the impurity atoms are substituted for the atoms of the host (this turns out a valid assumption) it is clear that, in chemists' language, the four covalent bonds to neighbouring atoms will leave one unused valence electron. Crudely we may imagine that this electron moves in the electrostatic potential of the impurity atom and that its ionisation energy may be calculated by analogy to the hydrogen atom. We know that the first ionisation energy of the hydrogen atom is 1 Rydberg (13·6 eV), but we may assume that this value will be reduced by substituting the effective mass of the electron and the dielectric constant of the host crystal. This approach leads to respective ionisation energies of 0·02 eV and 0·006 eV in silicon and germanium, so that the electrons may be described by discrete one-electron states at these energies below the bottom of the conductance band. Now just as an *electron* may be associated with the Group 5 atoms in silicon or germanium, we may associate *holes* with the Group 3 atoms boron or aluminium. The calculation of the energy of the hole states is analogous to that of the electron states; the energies (above the top of the valence band) would be identical if the effective masses were equal. In Table 6.2, we present

TABLE 6.2 Experimental electron and hole ionisation energies (eV) in the silicon and germanium (after Kittel)

Host	P	As	B	Al
Si	0·045	0·049	0·045	0·057
Ge	0·0120	0·0127	0·0104	0·0102

some experimental values of electron and hole ionisation energies in mild if accidental agreement with our calculations. We note that the values are well within the range of $k_B T$ at room temperature, so that equilibrium thermal excitation of the electrons and holes is expected. The levels associated with the Group 5 atoms are thus called (electron) *donor levels* and those associated with Group 3 atoms are called (electron) *acceptor levels*; the impurity atoms are called respectively *donor atoms* and *acceptor atoms*. Diagramatically, it is normal to indicate the donor and acceptor levels by dashed lines (see Fig. 6.16). A semiconductor whose conductivity depends on donor impurities is known as an *n*-type semiconductor and one whose conductivity depends on acceptor impurities is known as a *p*-type semiconductor.

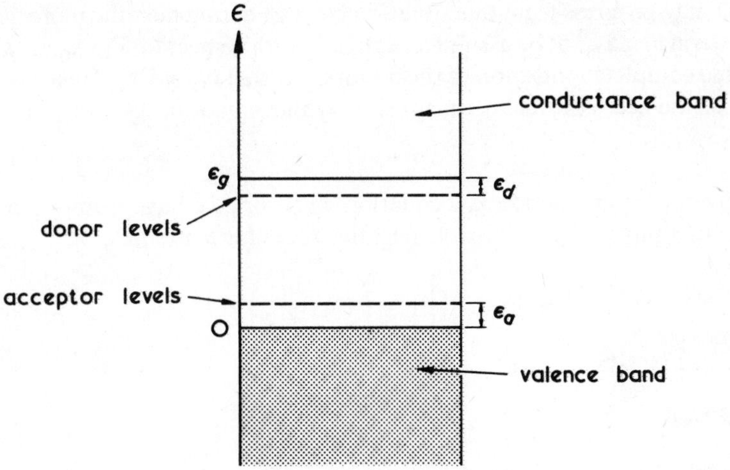

FIG. 6.16 Donor and acceptor levels.

Thermal ionisation of impurities

As a simple introductory example, let us consider an n-type semiconductor with N_d donor levels per unit volume, which lie just below the bottom of the conductance band, the impurity ionisation energy ε_d being very small compared with ε_g. If we assume that all of the donor atoms are ionised and that $N_d \gg n_i$, we may neglect the presence of intrinsic electrons in the conduction band. Thus we have the simple relationship

$$n = N_d \tag{6.172}$$

In this situation, n will be practically independent of temperature and we have what is called the *saturated extrinsic condition.* From eqns 6.167 and 6.172, we may write

$$N_d = \mathcal{N}_c \exp\left(\frac{\mu - \varepsilon_g}{k_B T}\right) \tag{6.173}$$

or

$$\mu = \varepsilon_g + k_B T \ln\left(\frac{N_d}{\mathcal{N}_c}\right) \tag{6.174}$$

It is clear from 6.174 that as N_d increases, the Fermi level will rise from its mid-gap 'intrinsic' position to one close to the bottom of the conduction band when $N_d \simeq \mathcal{N}_c$. When $N_d > \mathcal{N}_c$, the electrons in the conductance band may no longer be treated by the classical limit of the Fermi–Dirac function used in deriving 6.167.

We may progress from this situation by considering next the more general case when n_i may not be assumed negligible with respect to N_d. Again we will assume complete ionisation of the donors, but that $N_d < \mathcal{N}_c$. Hence we may express the necessary charge neutrality by the equation

$$p + N_d = n \tag{6.175}$$

But from the mass action law equation, $np = n_i^2$, we have from 6.175 (here neglecting physically impossible negative roots for n and p).

$$n = \frac{N_d}{2}\left[1 + \left(1 + \frac{4n_i^2}{N_d^2}\right)^{1/2}\right] \tag{6.176}$$

and

$$p = \frac{2n_i^2}{N_d}\left[1 + \left(1 + \frac{4n_i^2}{N_d^2}\right)^{1/2}\right]^{-1} \tag{6.177}$$

Arising from eqns 6.176 and 6.177, we may consider two possibilities depending upon $n_i \ll N_d$ or $n_i \gg N_d$. In the former case, we may carry out a binomial expansion of the square roots in 6.176 and 6.177 to the second term, leading to the equations

$$n = N_d + \frac{n_i^2}{N_d} \tag{6.178}$$

and

$$p = \frac{n_i^2}{N_d} \tag{6.179}$$

It will be observed that the number of holes (as expressed in 6.179) is much less than p_i since we may write instead of 6.179

$$p = p_i\left(\frac{n_i}{N_d}\right) \tag{6.180}$$

The holes may be referred to therefore as the *minority carriers* and the electrons as the *majority carriers*. If, on the other hand, $n_i \gg N_d$, we have

$$n = n_i + \frac{N_d}{2} \tag{6.181}$$

and

$$p = n_1 - \frac{N_d}{2} \tag{6.182}$$

As the next possibility, let us suppose that we have N_a acceptor levels per unit volume lying just above the top of the valence band. We shall assume that these levels are completely occupied by electrons from the valence band and that the intrinsic contribution is negligible. Hence from 6.168

$$\mu = k_B T \ln\left(\frac{\mathcal{N}_v}{N_a}\right) \tag{6.183}$$

an equation analogous to 6.174, which shows the decrease of μ towards the top of the valence band with increasing acceptor concentration N_a. Now following the same lines of argument used in the donor case, if we now consider the intrinsic contribution we have (and this should be checked by using the mass action equation and the neutrality condition)

$$n = \frac{2n_i^2}{N_a}\left[1 + \left(1 + \frac{4n_i^2}{N_a^2}\right)^{1/2}\right]^{-1} \tag{6.184}$$

and

$$p = \frac{N_a}{2}\left[1 + \left(1 + \frac{4n_i^2}{N_a^2}\right)^{1/2}\right] \tag{6.185}$$

So for $n_i \ll N_a$

$$n = \frac{n_i^2}{N_a} \tag{6.186}$$

and

$$p = N_a + \frac{n_i^2}{N_a} \tag{6.187}$$

and for $n_i \gg N_a$

$$n = n_i - \frac{N_a}{2} \tag{6.188}$$

and

$$p = n_i + \frac{N_a}{2} \tag{6.189}$$

We may now begin to consider the possibility of a semiconductor containing *both* donor and acceptor impurities. Let us suppose that n_i may be neglected by imposing the condition that $|N_d - N_a| \gg n_i$. When $N_a = 0$, eqn. 6.174 applies and the Fermi level will be appreciably above the middle of the band gap. As N_a is allowed to increase, there will be initially a very high probability of electrons occupying the acceptor levels at the expense of the donor levels and the conductance band. The number of *effective* donor levels will thus be $N_d - N_a$. Hence the position of the Fermi level may be considered to be determined by the equation

$$N_d - N_a = \mathcal{N}_c \exp\left(\frac{\mu - \varepsilon_g}{k_B T}\right) \tag{6.190}$$

An N_a approaches N_d, there will be a tendency for the effects of the acceptor and donor levels to cancel (crudely all the electrons from the latter will occupy the former) so that the Fermi level should resort to its intrinsic position. Analogously we may argue when $N_a > N_d$ and the total situation is summarised in Fig. 6.17.

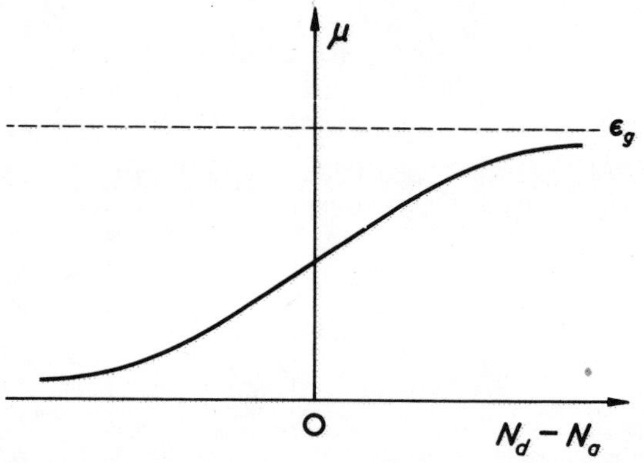

FIG. 6.17 Influence of donor and acceptor concentrations on the Fermi level.

So far we have assumed complete ionisation of the donor and acceptor impurities, but it is important to consider the consequences of incomplete ionisation at low temperatures. In order to treat this situation, we must consider the degeneracy of the impurity levels. Two conditions are seen likely to arise.

(1) An impurity level which can be empty or which may accept *one* electron of either spin.
(2) An impurity level which may contain two paired electrons or one of either spin.

The normal Fermi–Dirac function is clearly not appropriate under these conditions (see for example A. H. Wilson) and the modified function is

$$f(\varepsilon) = \frac{1}{g \exp\left(\frac{\varepsilon - \mu}{k_B T}\right) + 1} \tag{6.191}$$

The distribution function for the occupancy of a level by an electron of either spin (see condition 1) is given by eqn. 6.191 with $g = \frac{1}{2}$. For example, the occupation statistics for the odd valence electron of a Group 5 donor impurity in a Group 4 semiconductor would be

$$f(\varepsilon) = \frac{1}{\frac{1}{2}\exp\left(\frac{\varepsilon - \mu}{k_B T}\right) + 1} \tag{6.192}$$

so that if n_d is the concentration of unionised donors, we may write

$$n_d = \frac{N_d}{\frac{1}{2}\exp\left(\frac{\varepsilon_g - \varepsilon_d - \mu}{k_B T}\right) + 1} \tag{6.193}$$

(with reference to Fig. 6.16)

Similarly, the distribution function for the occupation of a level by two paired electrons (see condition 2) is given by eqn. 6.191 with $g = 2$. Thus the occupation statistics for a fourth electron of a Group 3 acceptor impurity in a Group 4 semiconductor would be

$$f(\varepsilon) = \frac{1}{2\exp\left(\frac{\varepsilon - \mu}{k_B T}\right) + 1} \tag{6.194}$$

Thus the concentration of unionised acceptors will be given by

$$n_a = N_a\left[1 - \frac{1}{2\exp\left(\frac{\varepsilon_a - \mu}{k_B T}\right) + 1}\right]$$

$$= \frac{N_a}{\frac{1}{2}\exp\left(\frac{\mu - \varepsilon_a}{k_B T}\right) + 1} \tag{6.195}$$

In order to calculate n and p, we start by expressing the condition for electrical neutrality

$$n + (N_a - n_a) = p + (N_d - n_d) \tag{6.196}$$

and by means of eqns 6.167, 6.168, 6.193, 6.195 and 6.196 we should, in principle, be able to solve for μ. For many real situations it is necessary to resort to numerical methods of solution, but for our purposes we may obtain useful information by making certain assumptions about the concentration of carriers. We will suppose that $N_d > N_a$ so that the majority carriers are electrons and we will asume that all of the acceptors are ionised i.e. the Fermi level is well above the bottom of the band gap. Thus eqn. 6.196 may be reduced to the simplified form

$$n + N_a = p + (N_d - n_d) \tag{6.197}$$

We may substitute for n_d in 6.197 by use of 6.193 and for p by use of the mass action relationship. Thus we have

$$n + N_a = \frac{n_i^2}{n} + N_d - \frac{N_d}{\frac{1}{2}\exp\left(\frac{\varepsilon_g - \varepsilon_d - \mu}{k_B T}\right) + 1} \tag{6.198}$$

Further, we may eliminate μ from 6.198 by means of eqn. 6.167 leading finally (and this should be checked) to the equation

$$\frac{n(n + N_a) - n_i^2}{N_d - N_a - n + n_i^2/n} = \frac{\mathcal{N}_c}{2}\exp\left(\frac{-\varepsilon_d}{k_B T}\right) \tag{6.199}$$

At very low temperatures it is reasonable to assume the following relative orders: $n \ll N_a$ and $n_i \ll n \ll (N_d - N_a)$

Hence eqn. 6.199 reduces to the form

$$\frac{nN_a}{(N_d - N_a)} \simeq \frac{\mathcal{N}_c}{2}\exp\left(\frac{-\varepsilon_d}{k_B T}\right) \tag{6.200}$$

or

$$n \simeq \frac{\mathcal{N}_c}{2}\left(\frac{N_d - N_a}{N_a}\right)\exp\left(\frac{-\varepsilon_d}{k_B T}\right) \tag{6.201}$$

From 6.201 it follows that

$$\mu \simeq \varepsilon_g - \varepsilon_d + k_B T \ln\left(\frac{N_d - N_a}{N_a}\right) \tag{6.202}$$

so we see that at low temperatures, the Fermi level is above the energy of the donor levels, to which it tends as the temperature $T \to 0$.

At a slightly higher temperature (the *reserve* region), we may assume still that $n_i \ll n \ll (N_d - N_a)$ but that $n \gg n_a$. Hence from 6.199

$$\frac{n^2}{N_d - N_a} \simeq \frac{\mathcal{N}_c}{2}\exp\left(\frac{-\varepsilon_d}{k_B T}\right) \tag{6.203}$$

or

$$n \simeq \left[\frac{\mathcal{N}_c(N_d - N_a)}{2}\right]^{1/2}\exp\left(\frac{-\varepsilon_d}{2k_B T}\right) \tag{6.204}$$

and

$$\mu \simeq \varepsilon_g - \frac{\varepsilon_d}{2} + \frac{k_B T}{2} \ln\left(\frac{N_d - N_a}{\mathcal{N}_c}\right) \tag{6.205}$$

At a still higher temperature, we may assume that all of the donors are ionised but still that $n \gg n_i$. Under these conditions we may write (cf. eqn. 6.172)

$$n \simeq N_d - N_a \tag{6.206}$$

and

$$\mu = \varepsilon_g + k_B T \ln\left(\frac{N_d - N_a}{\mathcal{N}_c}\right) \tag{6.207}$$

As before we label that region of temperature in which 6.206 applies, the *saturation* (or *exhaustion*) region. Finally we may define an intrinsic region in which $n \simeq n_i$. Here we may use 6.167 for n and 6.170 for μ. The values of n and μ as a function of temperature are summarised in the familiar diagrams of Figs 6.18 and 6.19 and the form of the curves when $N_a > N_d$ will be self-evident. In practical devices, it is usually necessary to operate in the extrinsic or exhaustion regions so that the temperature at which intrinsic behaviour begins to predominate may dictate the maximum working temperature of the device. With a higher value of ε_g it is thus possible to operate silicon semiconductors at a higher temperature than germanium semiconductors.

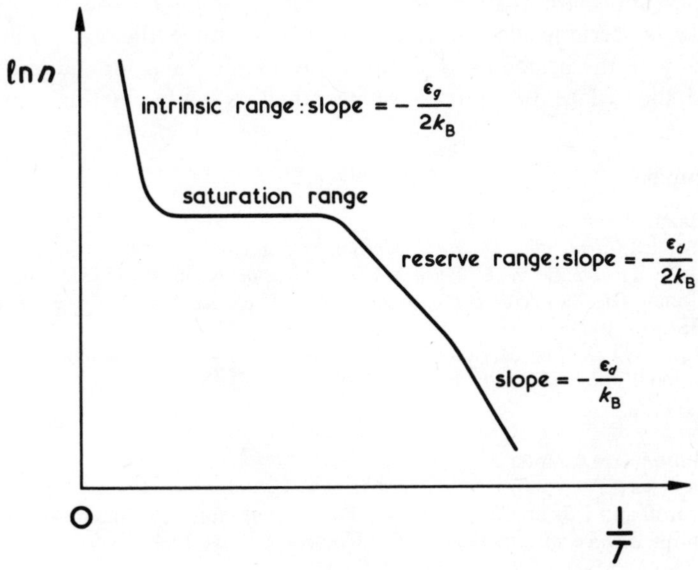

FIG. 6.18 Temperature dependence of semiconductor behaviour.

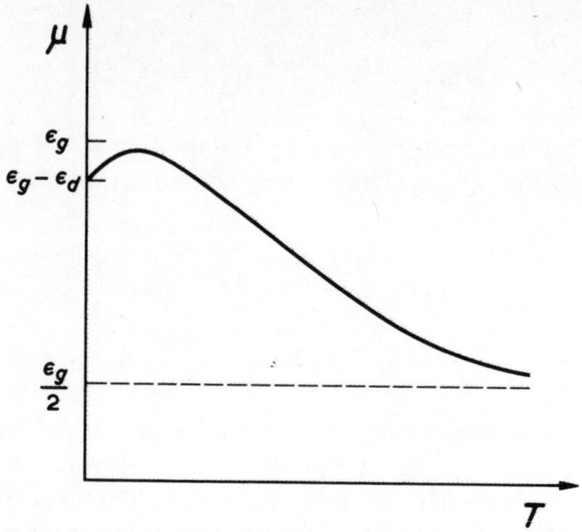

FIG. 6.19 Effect of temperature on the position of the Fermi level in a semiconductor with electrons as majority carriers.

Although our discussion of semiconductor physics has been very brief and elementary, the reader should be able to tackle more advanced topics without too much difficulty. For example, it might be a good idea to consider the interesting question of the transport properties of semiconductors (certainly information about the Hall effect should be sought). Further enquiries concerning effective mass would be desirable (the reader might get to grips with the principles of cyclotron resonance) and for the practically minded, the surface properties of semiconductors is a fascinating topic.

References

(1) F. Bloch, *Z. Physik* **52**, 555 (1928)
(2) M. Scharff, *Elementary Quantum Mechanics* (John Wiley, 1969)
(3) R. de L. Kronig and W. C. Penney, *Proc. Roy. Soc.* **A130**, 499 (1931)
(4) H. Jones, *The Theory of Brillouin Zones and Electronic States in Crystals* (North Holland, 1960)
(5) S. Raimes, *The Wave Mechanics of Electrons in Metals* (North Holland, 1961)
(6) E. H. Hall, *Amer. J. Math.* **2**, 287 (1879)

General

A. H. Wilson, (*see* Chapter 5)
S. Raimes (*see* (5))
L. V. Azaroff and J. J. Brophy, *Electronic Processes in Materials* (McGraw-Hill, 1963)
R. A. Smith, *Semiconductors* (Cambridge University Press, 1968)

7

Imperfections

In our discussions we have assumed (and indeed emphasised) that the crystalline state involves perfect periodicity. Real crystals, however, are not endowed with this perfection, but as a bonus many interesting and important properties of crystals depend upon their containing *imperfections*. Imperfections (or defects) may be defined generally as any structural element which results in a deviation from crystal periodicity. Thus, for example, we consider any impurity in a solid as constituting an imperfection.

Imperfections are normally classified according to their *dimension* and we will deal with those of zero dimension (point defects) and unit dimension (line defects).

7.1 Point defects

As its name implies, this defect involves single atoms and in Fig. 7.1 we represent possible defects of this type with reference to a simple ionic solid. The simplest point defect is a *lattice vacancy* which may be considered formally as resulting from the transfer of an atom from a lattice site in the bulk to the surface of the crystal. This is known as a *Schottky defect* and at any finite temperature an otherwise perfect crystal must contain an equilibrium concentration of these as determined by simple statistical considerations of configurational entropy.

According to classical thermodynamics, the equilibrium condition of a solid of constant volume at a temperature T is the minimisation of the free energy $F(=U - TS)$. We let E_{vac} equal the energy required to remove an atom from its normal lattice site to a surface site. Thus if we assume that the vacancies do not interact, the energy of formation of n vacancies is nE_{vac}. The entropy change in creating n vacancies in a total of $N + n$ lattice sites may be considered to be of two types:

(1) The vibrational entropy contribution
(2) The configurational entropy contribution.

The exact form of (1) will not concern us particularly (as its total contribution is small, but we could, for example, use the Einstein model to appreciate its form). The configurational entropy term, on the other hand, may be obtained

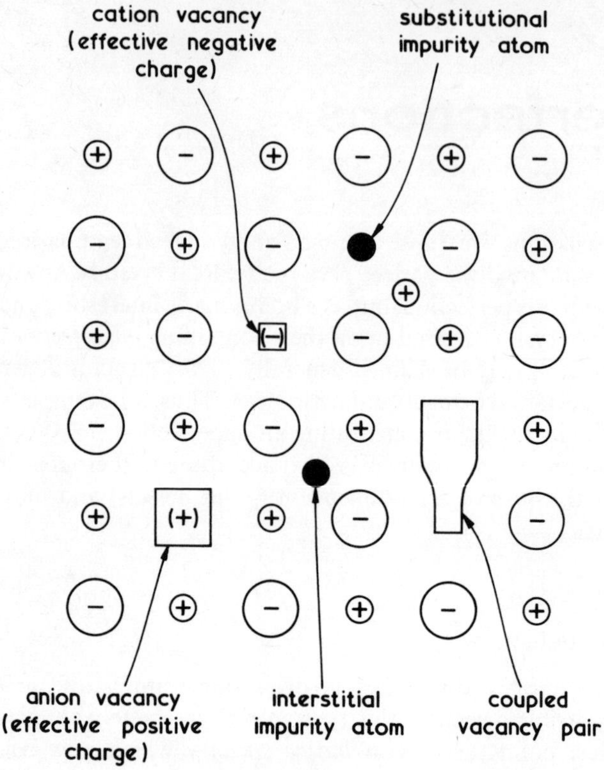

FIG. 7.1 Point defects in an ionic solid.

directly from the Boltzmann equation

$$S = k_B \ln \Omega \tag{7.1}$$

$$= k_B \ln\left[\frac{(N+n)!}{N!n!}\right] \tag{7.2}$$

Hence we may write

$$F = nE_{\text{vac}} - k_B T \ln\left[\frac{(N+n)!}{N!n!}\right] \tag{7.3}$$

The factorials may be expanded using Stirling's approximation

$$\ln x! \simeq x \ln x - x$$

leading to

$$\ln\left[\frac{(N+n)!}{N!n!}\right] = (N+n)\ln(N+n) - N \ln N - n \ln n \tag{7.4}$$

Using 7.4 in 7.3 and differentiating, we obtain

$$\left(\frac{\partial F}{\partial n}\right)_{T,V} = E_{\text{vac}} - k_{\text{B}} T \ln\left(\frac{N+n}{n}\right) \tag{7.5}$$

At equilibrium, we require that $(\partial F/\partial n) = 0$, so that from 7.5:

$$E_{\text{vac}} = k_{\text{B}} T \ln\left(\frac{N+n}{n}\right) \tag{7.6}$$

For $n \ll N$, eqn. 7.6 may be written in the simpler form

$$n \simeq N \exp\left(-\frac{E_{\text{vac}}}{k_{\text{B}} T}\right) \tag{7.7}$$

which indicates that the equilibrium concentration of defects n/N is determined solely by the energy of formation and the absolute temperature (see Table 7.1).

TABLE 7.1 Equilibrium concentration of Schottky defects (using $E_{\text{vac}} = 1$ eV)

T(K)	$\sim n/N$
200	$6{\cdot}5 \times 10^{-26}$
400	$2{\cdot}5 \times 10^{-13}$
600	$4{\cdot}2 \times 10^{-9}$
800	$5{\cdot}0 \times 10^{-7}$
1000	$9{\cdot}2 \times 10^{-6}$

In ionic crystals it would be expected, on general energetic grounds, that the formation of equal numbers of anion and cation vacancies would be favoured: thus local electroneutrality would be preserved. If E_{pair} is the energy required to form a pair of Schottky defects, an argument entirely analogous to that above leads to the expression

$$n \simeq N \exp\left(\frac{-E_{\text{pair}}}{2k_{\text{B}} T}\right) \tag{7.8}$$

where n is the number of vacancy pairs.

(Note the factor 2 in the denominator of the exponent; the reasons for its presence if not immediately obvious should be sought.)

A type of vacancy related to the Schottky defect is known as the *Frenkel defect*, which may be considered to result from the migration of an atom from a lattice site to an interestitial position (i.e. a position between lattice sites

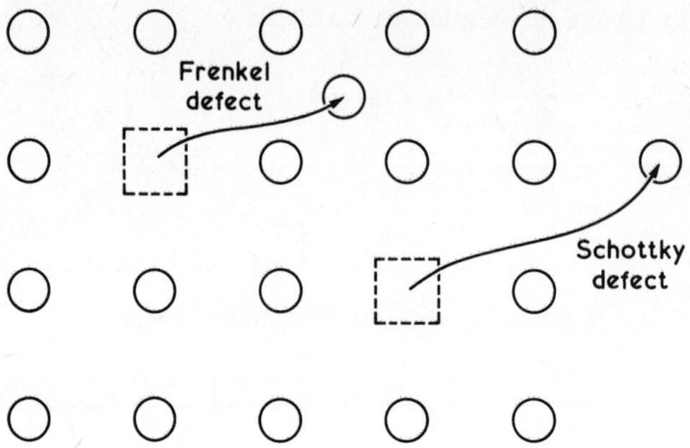

FIG. 7.2 The formation of Schottky and Frenkel defects.

and thus normally unoccupied). A pictorial representation of the formation of Schottky and Frenkel defects is given in Fig. 7.2. We leave it to the interested reader to consider a calculation of the equilibrium concentration of Frenkel defects.

One might ask the reasonable question: how do we know that vacancies occur in crystalline solids? The most direct approach is obviously a comparison of pyknometric and diffraction-based densities. Such measurements are difficult and often at the limit of experimental feasibility. However, the low density of pure alkali metal halides compared with that obtained from X-ray data indicates the predominance of Schottky defects. In contradistinction, the existence of Frenkel defects does not alter the pyknometric density to a first approximation. Hence if independent evidence for vacancies exists, agreement with density obtained from diffraction data is usually taken as good evidence for Frenkel defects. For example, pure silver halides appear to contain only this type of defect.

Important information about vacancies may be derived also from electrical conductivity, which we will consider in a subsequent section. But it is appropriate before a specific discussion of this topic to consider the problem of diffusion in solids. If a crystal were perfect (i.e. no vacancies or other defects) it would appear difficult to visualise the movement of ions. However, if studies of *self-diffusion* in crystals are made using radioactive tracers e.g. using ^{24}Na in a crystal of sodium chloride, it is found that diffusion occurs at a finite rate. In our example, it is observed that the activation energy for diffusion is 173·7 kJ mol^{-1} and it is supposed that the process occurs by way of vacant sites due to Schottky defects. We may suppose therefore that the activation energy is made up of two parts: a contribution which may be associated with the formation of a vacancy and a contribution from the energy required to move a Na^+ ion to an adjoining vacant site. Since this diffusion by the vacancy mechanism is a random process, we may treat it as a 'random walk problem'. This approach is widely used in other physical contexts (e.g. polymer statistics) so we deal with in a simple but fairly detailed way.

Random walk

Consider a particle (an atom or a vacancy) starting at a chosen origin and making n jumps, each represented by a vector $\boldsymbol{r}$ (see Fig. 7.3). The final position after n jumps may be represented by the vector $\boldsymbol{R}_n$. We see from Fig. 7.3 that

$$\boldsymbol{R}_n = \boldsymbol{r}_1 + \boldsymbol{r}_2 + \boldsymbol{r}_3 + \cdots + \boldsymbol{r}_{n-1} + \boldsymbol{r}_n \tag{7.9}$$

$$= \sum_{i=1}^{n} \boldsymbol{r}_i \tag{7.10}$$

To obtain the magnitude R_n we may square each side of 7.9 leading to

$$\begin{aligned} R_n^{\,2} = {} & \boldsymbol{r}_1\cdot\boldsymbol{r}_1 + \boldsymbol{r}_1\cdot\boldsymbol{r}_2 + \boldsymbol{r}_1\cdot\boldsymbol{r}_3 + \cdots + \boldsymbol{r}_1\cdot\boldsymbol{r}_n \\ & + \boldsymbol{r}_2\cdot\boldsymbol{r}_1 + \boldsymbol{r}_2\cdot\boldsymbol{r}_2 + \boldsymbol{r}_2\cdot\boldsymbol{r}_3 + \cdots + \boldsymbol{r}_2\cdot\boldsymbol{r}_n \\ & \quad \vdots \qquad\qquad \vdots \qquad\qquad \vdots \qquad\qquad\qquad \vdots \\ & + \boldsymbol{r}_n\cdot\boldsymbol{r}_1 + \boldsymbol{r}_n\cdot\boldsymbol{r}_2 + \boldsymbol{r}_n\cdot\boldsymbol{r}_3 + \cdots + \boldsymbol{r}_n\cdot\boldsymbol{r}_n \end{aligned} \tag{7.11}$$

$$= \sum_{i=1}^{n} \boldsymbol{r}_i\cdot\boldsymbol{r}_i + 2\left\{\begin{array}{r} \boldsymbol{r}_1\cdot\boldsymbol{r}_2 + \boldsymbol{r}_1\cdot\boldsymbol{r}_3 + \boldsymbol{r}_1\cdot\boldsymbol{r}_4 \ldots \boldsymbol{r}_1\cdot\boldsymbol{r}_n \\ + \boldsymbol{r}_2\cdot\boldsymbol{r}_3 + \boldsymbol{r}_2\cdot\boldsymbol{r}_4 \ldots \boldsymbol{r}_2\cdot\boldsymbol{r}_n \\ \vdots \\ \boldsymbol{r}_{n-1}\cdot\boldsymbol{r}_n \end{array}\right\} \tag{7.12}$$

$$= \sum_{i=1}^{n} \boldsymbol{r}_i\cdot\boldsymbol{r}_i + 2\sum_{j=1}^{n-1}\sum_{i=1}^{n-j} \boldsymbol{r}_i\cdot\boldsymbol{r}_{i+j} \tag{7.13}$$

From elementary vector algebra we know that

$$\boldsymbol{r}_i\cdot\boldsymbol{r}_{i+j} = r_i r_{i+j}\cos\theta_{i,i+j} \tag{7.14}$$

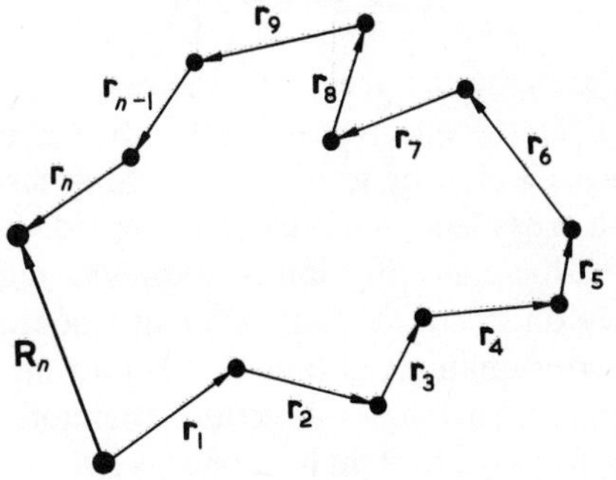

FIG. 7.3 The random walk.

where θ is the angle between vector $\boldsymbol{r}_i$ and vector $\boldsymbol{r}_{i+j}$.

Thus

$$R_n^2 = \sum_{i=1}^{n} r_i^2 + 2\sum_{j=1}^{n-1} \sum_{i=1}^{n-j} r_i r_{i+j} \cos \theta_{i,i+j} \tag{7.15}$$

We may simplify eqn. 7.15 by considering atoms or vacancies in crystals with cubic symmetry so that the magnitude of the vectors will be equal (to r). Hence

$$R_n^2 = nr^2\left(1 + \frac{2}{n}\sum_{j=1}^{n-1}\sum_{i=1}^{n-j} \cos \theta_{i,i+j}\right) \tag{7.16}$$

The quantity R_n^2 is appropriate to one particle after n jumps. An average value of R_n^2 ($\overline{R_n^2}$) can be found by considering a number of particles each of which will have undergone n jumps.

$$\overline{R_n^2} = nr^2\left(1 + \frac{2}{n}\sum_{j=1}^{n-1}\sum_{i=1}^{n-j} \overline{\cos \theta_{i,i+j}}\right) \tag{7.17}$$

If each jump direction is random, then positive and negative values of any $\cos\theta$ will occur with equal probability. Hence the double summation of eqn. 7.17 may be equated to zero, so that

$$\overline{R_n^2} = nr^2 \tag{7.18}$$

or

$$(\overline{R_n^2})^{1/2} = n^{1/2}r \tag{7.19}$$

We now wish to relate this equation to the so-called *diffusion coefficient* or *diffusivity D* of Fick's first diffusion law, which in its one-dimensional form may be written

$$J = -D\frac{\mathrm{d}c}{\mathrm{d}x} \tag{7.20}$$

Equation 7.20 relates the net flux J to the concentration gradient $\mathrm{d}c/\mathrm{d}x$ and we may consider the effect of the latter on the diffusive motion of a collection of particles. Referring to Fig. 7.4, we consider two regions of width $(\overline{R_n^2})^{1/2}$ and of unit cross sectional area on either side of a reference plane. If c at a time $t = 0$ is the concentration of species in region 1, then the total number in this region is $c(\overline{R_n^2})^{1/2}$. If one third of the particles diffuse in each of the three coordinate axial directions in a time t, then $\frac{1}{6}$ will diffuse in the $+x$ direction and $\frac{1}{6}$ in the $-x$ direction. Therefore, the number crossing the reference plane from left to right in a time t is $\frac{1}{6}(\overline{R_n^2})^{1/2}c$. If a concentration gradient $\mathrm{d}c/\mathrm{d}x$ exists in region 2, the average particle concentration at a

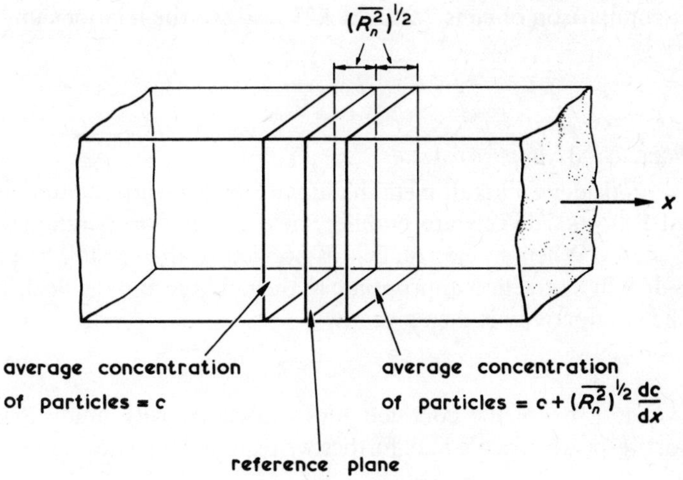

FIG. 7.4 One-dimensional diffusion.

time $t = 0$ will be

$$c + (\overline{R_n^2})^{1/2} \frac{\mathrm{d}c}{\mathrm{d}x}$$

so that the number crossing the reference plane from right to left in a time t will be

$$\frac{1}{6}(\overline{R_n^2})^{1/2} c + \frac{1}{6} \overline{R_n^2} \frac{\mathrm{d}c}{\mathrm{d}x}$$

Thus the net flux

$$J = -\frac{\overline{R_n^2}}{6t} \frac{\mathrm{d}c}{\mathrm{d}x} \tag{7.21}$$

Using 7.18 we may now substitute for $\overline{R_n^2}$ in 7.21 to give

$$J = -\frac{nr^2}{6t} \frac{\mathrm{d}c}{\mathrm{d}x}$$

We may define a jump frequency τ such that

$$\tau = \frac{n}{t} \tag{7.22}$$

Hence

$$J = -\frac{\tau r^2}{6} \frac{\mathrm{d}c}{\mathrm{d}x} \tag{7.23}$$

so that comparison of eqns 7.20 and 7.23 leads to the relationship

$$D = \frac{\tau r^2}{6} \tag{7.24}$$

Ionic conductivity in pure crystals

Let us consider an alkali metal halide in which positively and negatively charged Schottky defects are mobile. Since the motion is random, no net current flows. When an electric field is applied to the crystal, however, the defects drift in a direction appropriate to their charge and the field. The total observed conductivity is then given by

$$\sigma = \sigma_c + \sigma_a \tag{7.25}$$

where σ_a and σ_c are the conductivities associated with anion and cation transport respectively. We may further write, following eqn. 7.8

$$\begin{aligned} \sigma &= ne\mu \\ &= Ne(\mu_c + \mu_a)\exp\left(\frac{-E_{\text{pair}}}{2k_B T}\right) \end{aligned} \tag{7.26}$$

where μ_c and μ_a are the respective cation and anion mobilities. In the absence of the electric field, random migration of the vacancies occurs according to the self diffusion constant D which we have related to the jump frequency. The latter quantity may be equated to the atomic vibrational frequency (in the Einstein approximation) modified by the Boltzmann factor, $\exp(-E_{\text{mob}}/k_B T)$ where E_{mob} is the barrier to movement of an atom from its lattice site to an adjoining vacancy. Thus we may write, using 7.24

$$D = \frac{\omega r^2}{6}\exp\left(\frac{-E_{\text{mob}}}{k_B T}\right) \tag{7.27}$$

The applied electric field $\mathscr{E}$ lowers the barrier by a value λ so favouring ionic migration in that direction. Equation 7.27 may be modified, therefore, to account for this effect:

$$D = \frac{\omega r^2}{6}\exp - \left(\frac{E_{\text{mob}} - \lambda}{k_B T}\right) \tag{7.28}$$

The ion transport will set up a vacancy gradient $\mathrm{d}n/\mathrm{d}x$ which will be related to the ion flux by Fick's first law.

$$J = D\frac{\mathrm{d}n}{\mathrm{d}x} \tag{7.29}$$

and the equilibrium condition will be that the sum of the diffusion and field induced drift currents is equal to zero. Thus

$$ne\mu\mathscr{E} + eD\frac{dn}{dx} = 0$$

or

$$n = A\exp\left(\frac{-\mathscr{E}x\mu}{D}\right) \tag{7.30}$$

where A is a constant.

Since we may write also

$$n \propto \exp\left(-\frac{\mathscr{E}ex}{k_B T}\right) \tag{7.31}$$

comparison of exponents, yields the Nernst–Einstein equation:

$$\frac{\mu}{D} = \frac{e}{k_B T} \tag{7.32}$$

so that substitution from 7.32 in 7.28 leads to the equation

$$\mu = \frac{e\omega r^2}{6k_B T}\exp - \left(\frac{E_{mob} - \lambda}{k_B T}\right) \tag{7.33}$$

Under normal conditions we may assume that $\lambda \ll k_B T$ so that as an acceptable approximation we write

$$\mu = \frac{e\omega r^2}{6k_B T}\exp\left(\frac{-E_{mob}}{k_B T}\right) \tag{7.34}$$

and substitution in 7.26 leads to the following equation for σ

$$\sigma = \frac{Ne^2r^2}{6k_B T}\left[\omega_c \exp - \left(\frac{E^+_{mob} + E_{pair}/2}{k_B T}\right) + \omega_a \exp - \left(\frac{E^-_{mob} + E_{pair}/2}{k_B T}\right)\right] \tag{7.35}$$

where the new symbols have obvious definitions.

Equation 7.35 describes *intrinsic conductivity* since it is a consequence of the existence of *intrinsic defects* in the lattice. For the alkali halides since $\mu_c \gg \mu_a$ a simplified equation

$$\sigma = \frac{Ne^2r^2}{6k_B T}\exp - \left(\frac{E^+_{mob} + E_{pair}/2}{k_B T}\right) \tag{7.36}$$

may be used except at high temperature and an Arrhenius plot will give the total activation parameter (see[1] for example, where fuller details will be found).

As well as the existence of intrinsic point defects, we may note also the importance of the introduction of point defects by doping. For example, cation vacancies in potassium chloride result from the addition of small amounts of a divalent chloride such as that of calcium. Clearly, if a Ca^{2+} ion occupies a normal cation lattice site and the two Cl^- ions occupy normal anion lattice sites, a cation vacancy is created due to the equivalence of the number of cation and anion sites in the crystal. Doping is important since it allows the controlled introduction of vacancies in materials of technical importance (besides semiconductors). A language appropriate to doping reactions, commonly referred to as the Kröger–Vink nomenclature has become common and although we will not deal with it here, the interested reader may find details elsewhere (for example in Swalin[2]).

Colour centres

Although associated in the first instance uniquely with the alkali metal halides, this term is now used to describe all point defects which visibly colour insulating materials. However, it is still appropriate to concentrate our discussion on the former.

As noted in Fig. 7.1 and implied in the discussion of ionic conductivity, vacancies in ionic crystals carry effective charges. Thus, an anion vacancy in the alkali metal halides of effective positive charge can act as an electron trap to become what is known as an *F-centre*. Coloured crystals of the alkali metal halides may be prepared typically by heating them in the vapour of an alkali metal and it is particularly noteworthy that the visible absorption maxima are independent of the alkali metal vapour employed. This, of course, would be the expected behaviour if the electronic transition responsible for the colour is to be associated only with the trapped electron.

In principle, positive holes might be expected to be similarly trapped at cation vacancies although experimental evidence for this type of colour centre is not apparent. However, a related *self-trapped hole* or *V-centre*, which has been detected experimentally may be considered to be the result of the reaction between two adjacent X^- ions and a hole to give X_2^-. A number of other colour centres have been described but will not concern us; the curious reader will find described F_1, F'_1, F_2, F_3, V_1, V_2 ... etc. colour centres (see[2] for example).

Formally we might think of a *F-centre* as a very simple 1-electron problem and consider whether or not it might be possible to set up a simple quantitative model. For entertainment if for no other purpose, we might enquire about the consequences of considering the electron trapped in a box by walls of infinite barrier height. Using a cubic box of dimensions a, the normalised eigenfunctions are

$$\psi = \left(\frac{8}{a^3}\right)^{1/2} \sin\left(\frac{n_1 \pi x_1}{a}\right) \sin\left(\frac{n_2 \pi x_2}{a}\right) \sin\left(\frac{n_3 \pi x_3}{a}\right) \qquad (7.37)$$

The energy eigenvalues are

$$\varepsilon = \frac{\pi^2\hbar^2}{2ma^2}(n_1{}^2 + n_2{}^2 + n_3{}^2) \tag{7.38}$$

The state of lowest energy is thus described by the quantum numbers $n_1 = n_2 = n_3 = 1$, so that the ground state energy is given by

$$\varepsilon_1 = \frac{3\pi^2\hbar^2}{2ma^2} \tag{7.39}$$

The first excited state will have energy

$$\varepsilon_2 = \frac{3\pi^2\hbar^2}{ma^2} \tag{7.40}$$

so that the first electronic transition would be expected at an energy corresponding to

$$\varepsilon_2 - \varepsilon_1 = \frac{3\pi^2\hbar^2}{2ma^2} \tag{7.41}$$

Of course, this approach may be outrageously crude, but eqn. 7.41 does suggest an inverse square relationship between the energy of the absorption band and the parameter a. Naturally, we do not know the 'size' of the trapped electron a but let us assume it is at least directly related to the lattice parameter. The relationship is tested in Fig. 7.5, although we prefer not to comment; this is left to the reader!

A number of other aspects of point defects should not totally escape our notice, and will constitute useful independent work for the student of this

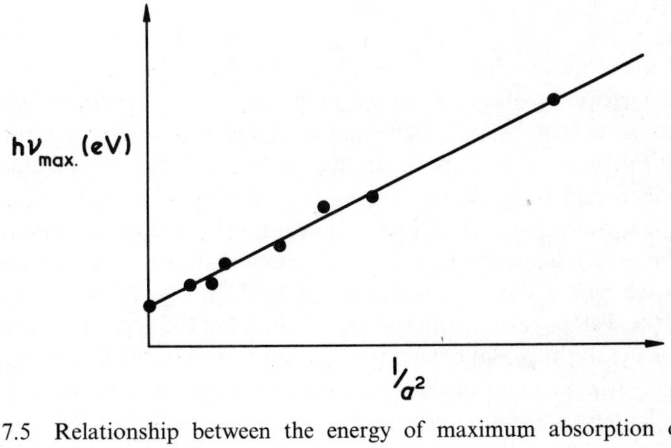

FIG. 7.5 Relationship between the energy of maximum absorption of colour centres and lattice parameters.

topic. For example, there is little need to stress the importance of alloys which may be thought of as solids containing gross point defects; also experimental aspects of point defects merit some attention.

7.2 Dislocations

Plastic deformation

So far in this chapter we have discussed imperfections defined by a point; as an introduction to imperfections defined by a line (commonly referred to as *dislocations*) it is apposite to consider a process known as *plastic deformation*, which may be observed experimentally for example, by applying a stress to a cylindrical single crystal of pure aluminium. The observed stress–strain relationship is qualitatively represented in Fig. 7.6. In the initial region,

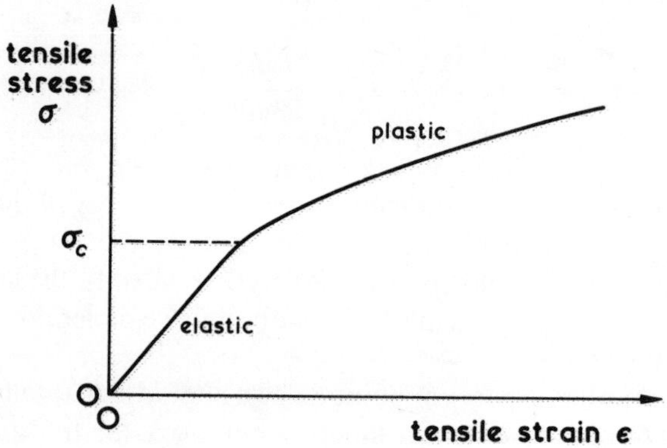

FIG. 7.6 Plastic deformation.

where Hook's law is obeyed [$\sigma = M\varepsilon$; (M = Young's modulus)] a *reversible* deformation is observed so we may label it *elastic*. Subsequent to a critical stress σ_c a *permanent* deformation known as *plastic deformation* is observed. Further, microscopic examination of the surface of the aluminium crystal will reveal that, arising during the process of plastic deformation, lines in the form of parallel ellipses run around the elongated crystal. These are known as *slip lines*. In fact close examination shows that the lines are in fact *steps* which are shown (wildly exaggerated) in Fig. 7.7. We may anticipate that these steps result from the slip of blocks of the crystal over so-called *slip-planes*. At the slip-plane we might consider simply the force required to effect shear displacement of one plane of atoms past its continguous plane. Such a calculation was first considered by Frenkel (see Fig. 7.8). For small elastic strains, the stress τ is related to the displacement x by the following

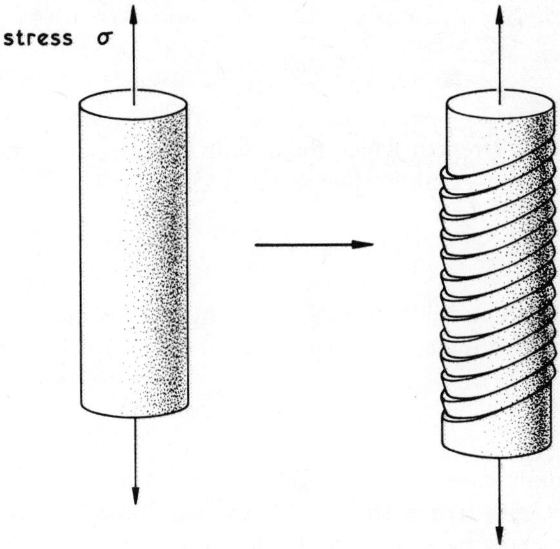

FIG. 7.7 Formation of slip lines under tensile stress.

equation

$$\tau = \frac{\mu x}{d} \tag{7.42}$$

where μ is the shear modulus. But for larger displacements, so that the atoms progress by a, we need another representation. As a first approximation we might represent the stress-displacement by a sine-function

$$\tau = k \sin \frac{2\pi x}{a} \tag{7.43}$$

The value of k may be evaluated by requiring that eqn. 7.43 must reduce to eqn. 7.42 for small displacements. Under these circumstances

$$\tau = \frac{\mu x}{d} = k \frac{2\pi x}{a} \tag{7.44}$$

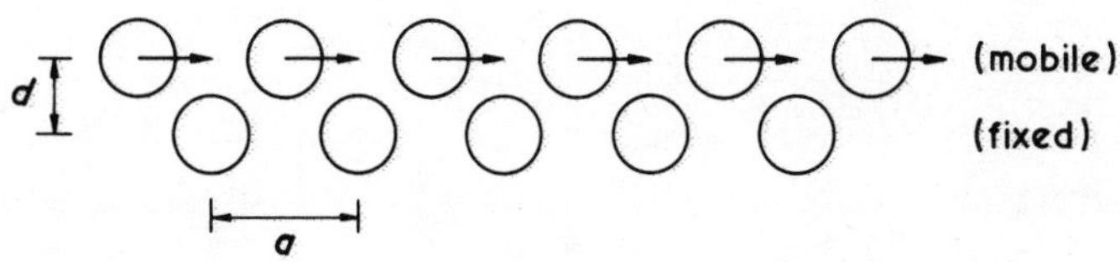

FIG. 7.8 Representation of shear displacement.

Therefore

$$\tau = \frac{\mu a}{2\pi d} \sin \frac{2\pi x}{a} \tag{7.45}$$

The critical shear stress at which the atomic layers become mobile is then assumed to be the maximum value of τ in 7.45. i.e.

$$\tau_c = \frac{\mu a}{2\pi d} \tag{7.46}$$

Since this is a rather crude approach anyway, no significant error will be introduced if we further approximate by letting $a = d$ in 7.46 to give

$$\tau_c = \frac{\mu}{2\pi} \tag{7.47}$$

Thus for aluminium, we may take $\mu = 3 \times 10^{10}$ Nm^{-2}, so that our calculated value for τ_c is $\simeq 5 \times 10^9$ Nm^{-2}. We may now ask how this value compares with experiment. In order to effect this comparison, it is necessary to resolve our tensile measurements into the appropriate shear component. This is readily done (Fig. 7.9) and we see that since

$$\tau = \sigma \cos \phi \cos \lambda \tag{7.48}$$

then

$$\tau_c = \sigma_c \cos \phi \cos \lambda \tag{7.49}$$

(τ_c is often called the critical shear stress for slip).

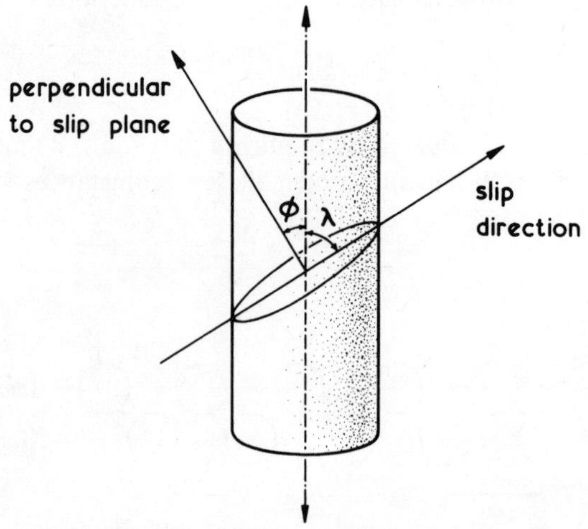

FIG. 7.9 Resolution of tensile stress.

From typical measured values of σ_c we find that $\tau_c \simeq 10^7$ Nm^{-2}, so there is a disparity of about three orders of magnitude between measured and calculated values of τ_c. This difference cannot be accounted for, even allowing for the crudeness of the chosen inter-atomic potential and even more realistically chosen potentials can only reduce τ_c to $\sim\mu/30$. So it is evident that the low value of shear strength can only be explained by reference to some other effects which act as sources of mechanical weakness. We appeal to dislocations as an aid to the resolution of this dilemma.

The suggestion that dislocations were responsible for facile slip in crystals, which was first made in 1934, has become firmly established. We have to consider two basic types of dislocation, edge and screw.

Edge and screw dislocations

Conceptually, the edge dislocation is the easier to appreciate and may, with respect to the model simple cubic lattice, be considered as the resultant of the inclusion of an extra plane of atoms terminating at the line *AB* in Fig. 7.10. Starting again from the *dislocation line AB*, we may introduce a screw dislocation by slip in the plane *ABCD*.

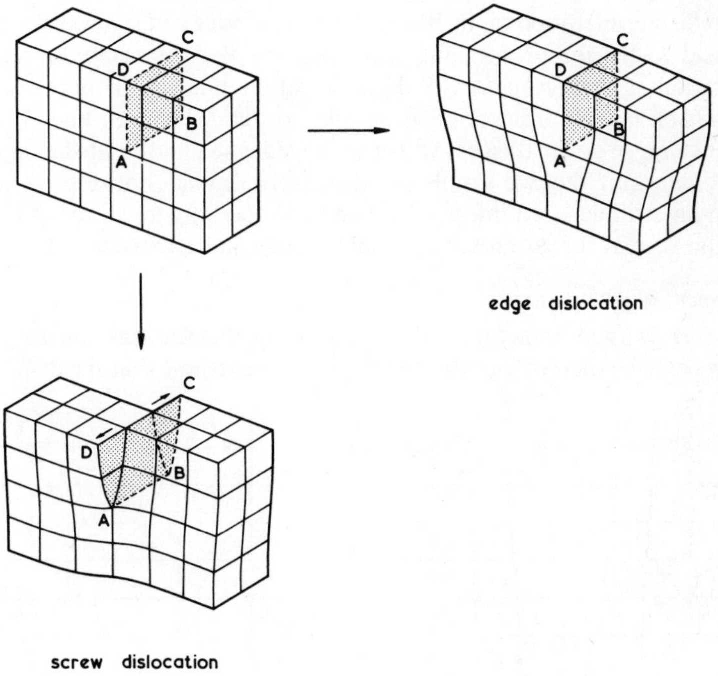

FIG. 7.10 Edge and screw dislocations.

The rationale of using the term *line defect* to describe screw and edge dislocations is apparent from the way we have chosen to introduce these dislocations with respect to the line *AB*. However, we still need more rigorous definitions of edge and screw dislocations and these are provided by reference to their *Burgers vectors*. Confusion may sometimes arise here and it is desirable to be consistent with one convention, otherwise the utility of the definition may disappear if one is concerned with the *sense* of the dislocation.

Burgers vectors

Consider the edge dislocation, a section through which is shown in Fig. 7.11(a). The point *L* marks the position at which the dislocation line perpendicularly cuts the section. We choose an arbitrary positive direction along the dislocation line *L* (positive into the page here) and a closed *Burgers circuit* is then constructed in a clockwise direction through the crystal ($S \rightarrow T \rightarrow U \rightarrow V \rightarrow S$) to include both 'perfect' and 'imperfect' sections. This circuit is then redrawn on a 'perfect' lattice (Fig. 7.11(b)) in the same clockwise direction. In the perfect lattice, the terminal position *F* is not coincident with *S* and, in fact the vector *FS* identifies the Burgers vector of the dislocation ***b***. This convention, which is now widely used is often referred to as the *FS/RH* (finish-start/right-hand screw direction) convention.

Careful consideration of the Burgers vectors of edge and screw dislocations will lead to the important conclusion that the Burgers vector of an edge dislocation is *perpendicular* to the dislocation line (as in our example) but that of a screw dislocation is *parallel* to the dislocation line. Thus by direct reference to the Burgers vector we have a much more satisfactory way of distinguishing the two simple line defects. In general, however, the dislocation line will lie at an intermediate angle to the Burgers vector of the dislocation so that the latter will have mixed edge and screw character.

Movement of dislocations

Having learned something of the nature of dislocations, we may now briefly describe their role in slip behaviour. If we assume that the dislocation

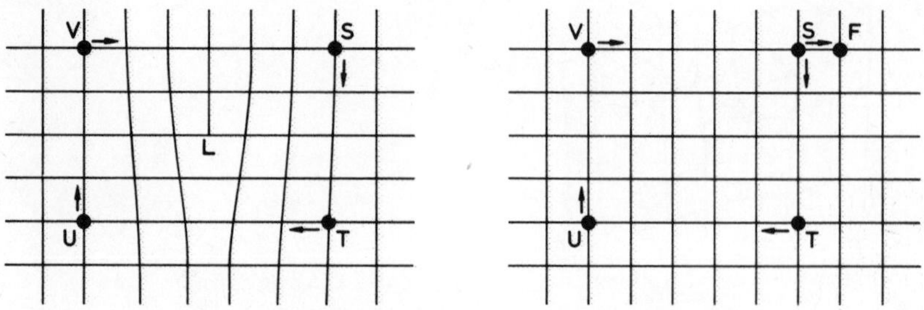

FIG. 7.11 The Burgers vector.

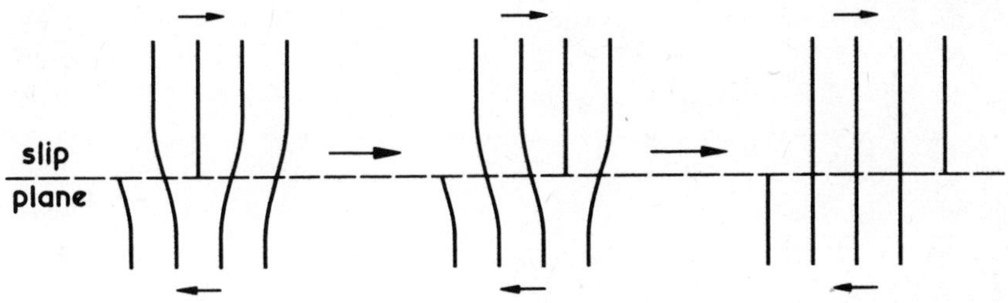

FIG. 7.12 Movement of an edge dislocation.

line of an edge dislocation and its Burgers vector lie in a convenient slip plane (although we have not discussed this point, slip may not occur in any plane or in any direction (see[3] for example)), the movement of the dislocation, giving effective slip, is shown in Fig. 7.12. The more facile glide compared with that expected in a perfect crystal is readily appreciated, since in contrast with the concerted movement in the latter, slip may be effected *piecemeal* with minor reorganisation around the dislocation line. The smaller resistance to movement is known as the Peierls–Nabarro force, details of which will not concern us here.

It will be noted that the edge dislocation line and its Burgers vector define a plane. If this plane is *not* a slip plane (see earlier note), the dislocation cannot move by this mechanism. Glide motion of a screw dislocation is less restricted, however, since the dislocation line and the Burgers vector (which are parallel) will lie in a number of planes. Hence a series of possible glide planes may be offered to the dislocation and, in fact, its motion may involve a number of them. This process is known as *cross-slip*.

Although this chapter has been short, the reader should not underestimate the importance of imperfections and extra reading is desirable. For example, defects may be involved in electron scattering processes in conductors, are very important in metallurgical processes (e.g. work hardening) and in chemical reactions.

References

(1) B. Henderson, *Defects in Crystalline Solids* (Edward Arnold, 1972)
(2) R. A. Swalin, *Thermodynamics of Solids* (John Wiley, 1972)
(3) D. Hull, *Introductions to Dislocations* (Pergamon Press, 1965)

General

D. Hull (*see* (3))
N. N. Greenwood, *Ionic Crystals, Lattice Defects and Non-stoichiometry* (Butterworth, 1968)
R. E. Reed-Hill, *Physical Metallurgy Principles* (Van Nostrand, 1964)
J. M. Thomas and W. J. Thomas, *Introduction to the Principles of Heterogeneous Catalysis* (Academic Press, 1967)

Index